Application of Soft Computing, Machine Learning, Deep Learning and Optimizations in Geoengineering and Geoscience

Wengang Zhang · Yanmei Zhang · Xin Gu ·
Chongzhi Wu · Liang Han

Application of Soft Computing, Machine Learning, Deep Learning and Optimizations in Geoengineering and Geoscience

Springer

Wengang Zhang
School of Civil Engineering
Chongqing University
Chongqing, China

Yanmei Zhang
College of Aerospace Engineering
Chongqing University
Chongqing, China

Xin Gu
School of Civil Engineering
Chongqing University
Chongqing, China

Chongzhi Wu
School of Civil Engineering
Chongqing University
Chongqing, China

Liang Han
School of Civil Engineering
Chongqing University
Chongqing, China

ISBN 978-981-16-6834-0 ISBN 978-981-16-6835-7 (eBook)
https://doi.org/10.1007/978-981-16-6835-7

This Springer imprint is published by the registered company Springer Nature Singapore Pte Ltd.
The registered company address is: 152 Beach Road, #21-01/04 Gateway East, Singapore 189721, Singapore

Preface

The so-called Fourth Paradigm has been boomingly developed during the past two decades, in which large quantities of observational data are available to scientists and engineers. Big data is characterized by the rule of the five Vs: Volume, Variety, Value, Velocity and Veracity. The concept of big data naturally matches well with the features of geoengineering and geoscience. Large-scale, comprehensive, multidirectional and multifield geotechnical data analysis is becoming a trend. On the other hand, soft computing (SC), machine learning (ML), deep learning (DL) and optimization algorithm (OA) provide the ability to learn from data and deliver in-depth insight into geotechnical problems. Researchers use different SC, ML, DL and OA models to solve various problems associated with geoengineering and geoscience. Consequently, there is a need to extend its research with big data research through integrating the use of SC, ML, DL and OA techniques.

This book focuses on the state of the art and application of SC, ML, DL and OA algorithms in geoengineering and geoscience. Various SC, ML, DL and OA approaches are firstly concisely introduced, concerning mainly the easy-to-interpret multivariate adaptive regression splines (MARS) model, supervised learning, unsupervised learning, deep learning and optimization algorithms. Then their representative applications in the geoengineering and geoscience are summarized via VOSviewer demonstration. The authors also provided their own thoughts learnt from these applications as well as work ongoing and future recommendations. This book aims to make a comprehensive summary and provide fundamental guidelines for researchers and engineers in discipline of geoengineering and geoscience or similar research areas on how to integrate and apply SC, ML, DL and OA methods.

Chongqing, China

Wengang Zhang
Yanmei Zhang
Xin Gu
Chongzhi Wu
Liang Han

Contents

About the Authors

Dr. Wengang Zhang is currently a full professor in School of Civil Engineering, Chongqing University, China. His research interests focus on impact assessment on the built environment induced by underground construction, as well as big data and machine learning in geotechnics and geoengineering. He is now the member of the ISSMGE TC304 (Reliability), TC309 (Machine Learning) and TC219 (System Performance of Geotechnical Structures). He has been selected as the World's Top 2% Scientists 2020.

Dr. Yanmei Zhang is currently an associate professor in College of Aerospace Engineering, Chongqing University, China. Her research interests focus on fatigue and failure assessment of engineering structures, fracture analyses of polymeric composites and applications of machine learning technique in geoengineering. She is the member of Chinese Society of Theoretical and Applied Mechanics.

Xin Gu is a Ph.D. candidate in the School of Civil Engineering, Chongqing University. He majors in geotechnical engineering, and his research interests involve random finite element analysis, slope stability analysis, spatial variability, Bayesian inference and machine learning.

Chongzhi Wu is currently Ph.D. student of Chongqing University, China. His research interest involves the application of machine learning methods in geotechnical engineering, and he is interested in time series analysis. He has experience on feature engineering, data mining and data visualization and has published several research papers about the application of ML methods in geotechnical engineering.

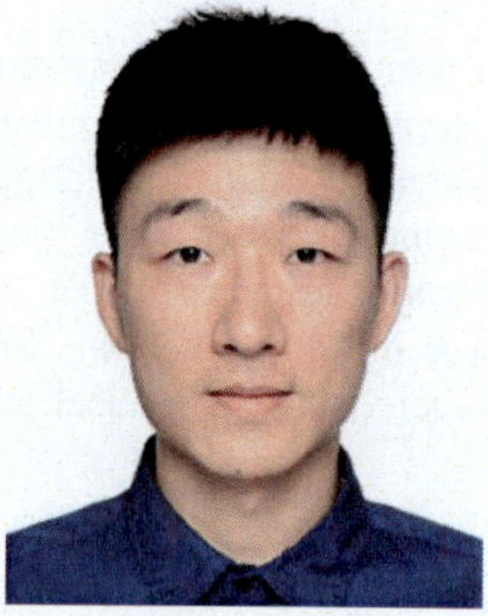

Liang Han is presently a Ph.D. candidate of Chongqing University, China. His research interest involves the statistical characterization of geomaterial parameters, site similarity identification and seismic performance evaluation of underground structures. He has experience on data analysis using Bayesian framework and machine learning. He has published several academic papers about the data analysis of geomaterial parameters and seismic performance of utility tunnels.

Abbreviations

ABC	Artificial bee colony
ACO	Ant colony optimization
AE	Automatic encoder
AI	Artificial intelligence
ANFIS	Adaptive neuro-fuzzy inference system
ANN	Artificial neural network
BC	Bayesian classifier
BF	Bacterial foraging
BM	Boosting methods
BO	Bayesian optimization
CG	Conjugate gradient
CNN	Convolutional neural network
CS	Cuckoo search
CSO	Cat swarm optimization
DBN	Deep belief network
DE	Differential evolution
DL	Deep learning
DT	Decision tree
ELM	Extreme learning machine
FA	Firefly algorithm
GA	Genetic algorithm
GEP	Gene expression programming
GP	Genetic programing
GRNN	Generalized regression neural network
GWO	Gray wolf optimizer
INSAR	Interferometric synthetic aperture radar
KNN	K-nearest neighbor
LiDAR	Light detection and ranging
LR	Linear regression
LSTM	Long short-term memory
M5	M5 model tree

Chapter 1
Introduction

Before introduction of more commonly used Artificial intelligence (AI), Machine learning (ML), Deep learning (DL) and Optimization algorithm (OA) technical expressions, the definition of Soft computing (SC) should be firstly mentioned since the former three terms are more relevant with each other.

SC is the use of approximate calculations to provide imprecise but usable solutions to complex computational problems. The approach enables solutions for problems that may be either unsolvable or just rather time-consuming to solve with current hardware. SC is sometimes referred to as computational intelligence, for comparison with the hard computing. It provides an approach to problem-solving using means other than computers. With the human mind as a role model, SC is tolerant of partial truths, uncertainty, imprecision and approximation, unlike traditional computing models. The tolerance of SC allows researchers to approach some problems that traditional computing can't process.

With the rapid development of computer performance, AI has gradually become a significantly vital driver for the fourth industrial revolution. The conception of AI was born at the Dartmouth conference in 1956 by several computational scientists (e.g., John McCarthy, Marvin Minsky, Claude Shannon) and their original idea was to discuss the usage of machines to mimic human learning and some other aspects of intelligence. Since 2012, the prompt increase of data volume induced by superior monitoring technology or devices (Dikshit et al. 2020; Gupta et al. 2020; Meng et al. 2021; Reichstein et al. 2019), the improvement of computing power and the emergence of a new ML algorithm, the DL have constantly contributed to the hot popularity of AI. AI has widely covered many fields, such as expert systems, machine learning, evolutionary computing, computer vision, natural language processing, etc. Although three commonly-seen terms "AI, ML and DL" have connections with each other, yet they are not the same indeed, which can be seen from Fig. 1.1 with highlighting the difference(Dikshit et al. 2020; Reichstein et al. 2019), also shown in Fig. 1.1 is SC. Generally speaking for ML, it can be considered as a method to realize AI, capable of capturing the internal patterns from data and then provide a rational decision as a guidance. Inspired by the way that human brains process information,

W. Zhang et al., *Application of Soft Computing, Machine Learning, Deep Learning and Optimizations in Geoengineering and Geoscience*,
https://doi.org/10.1007/978-981-16-6835-7_1

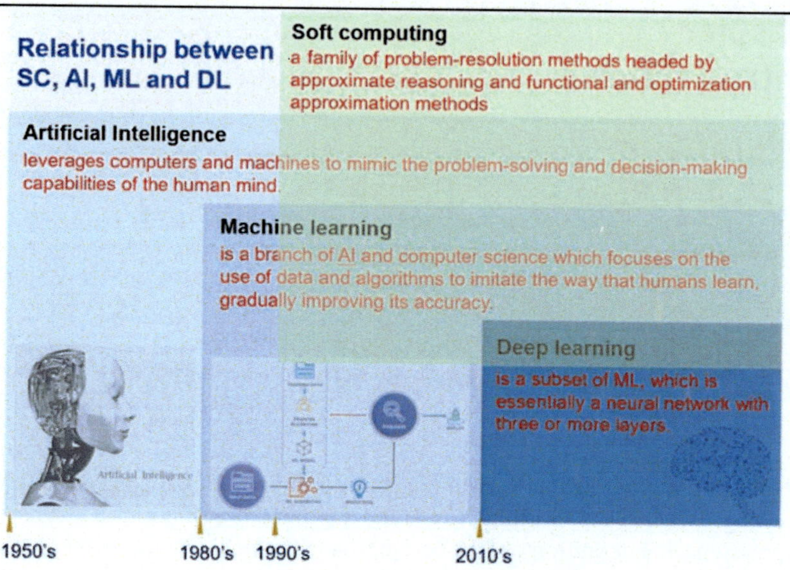

(a) Relationship between SC, AI, ML and DL

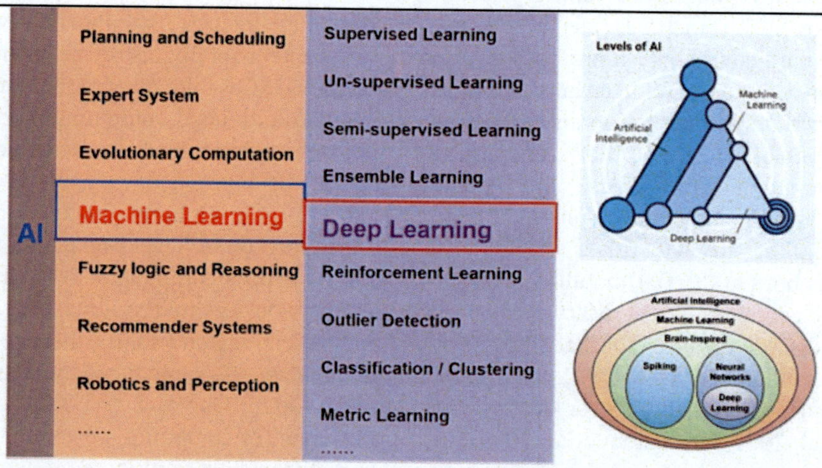

(b) Relation of inclusion for AI, ML and DL

Fig. 1.1 Different aspects of AI, ML, DL and SC

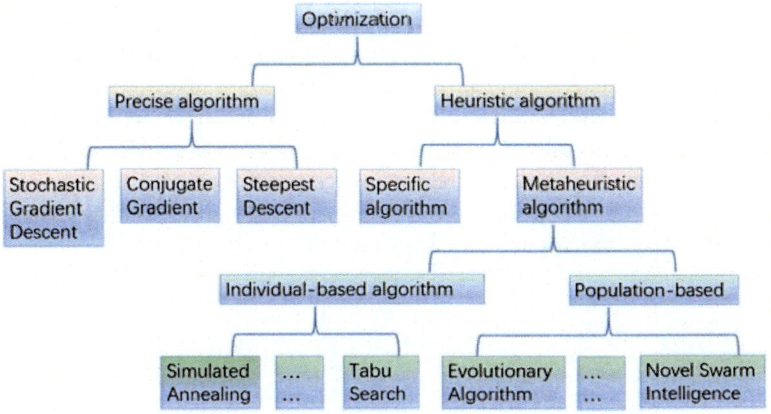

Fig. 1.2 Basic components of OA algorithms

DL is proposed as a main branch of ML, which always use more complex multi-layer neural network architectures (LeCun et al. 2015). Compared with other ML methods, DL requires less human guidance, while requires enormous amounts of data to explore complex, various, and inherent relationships hidden in data. OA denotes the optimization algorithms. The essence of most ML algorithms is to build a surrogate model and then optimize the objective function (or loss function) through OA, to obtain the optimal model with the best performance. Figure 1.2 plots the basic components of OA algorithms.

Currently, more and more attention has been focused on geohazards or the estimation of the safety of geotechnical structures considering that geohazards or failure of a structure would pose a great threat to the lives and property of the humans, such as Nicoll Highway Collapse, Singapore (Zhang et al. 2018); rockburst in Jinchuan 2nd Mine(Yi et al. 2010), Shenzhen landslide (Luo et al. 2019; Zhan et al. 2021); Bazimen landslide (Zhang et al. 2020b); Wenjia gully debris flow (Liu et al. 2021). To prevent or mitigate the disasters and ensure the geotechnical structures with sufficient safety, accurate predictions for occurrence of geohazards and reliability analysis for geotechnical structures are urgently needed.

In recent years, with the rapid development of SC, AI, ML, DL, and OA techniques, these methods are gradually widely adopted to conduct the assessment for the occurrence probability of geohazards and the destroy or deformation of geotechnical structures with enough accuracy and efficiency(Darabi et al. 2012; Li et al. 2014; Liang et al. 2012; Luo et al. 2019; Mohammadi et al. 2015; Neaupane and Adhikari 2006; Shi et al. 2019; Solomatine and Xue 2004; Zhan et al. 2021). Although there have been some high-level overviews of ML for the prediction of some specific areas in geotechnical engineering such as slopes, tunnels, deep braced excavations (Sheil et al. 2020; Shreyas and De 2019; Zhang et al. 2020a), yet a more comprehensive investigation of recent research progress and applications of SC, ML, DL and OA in geotechnical engineering and geoscience is quite limited. Therefore, this book first gives a brief introduction of typical SC, ML, DL and OA mainly via VOSviewer, a software tool for constructing and visualizing bibliometric networks, then some representative applications of these algorithms in geotechnical engineering and geoscience are systematically summarized. Furthermore, reflections from these applications, ongoing research and the future potentials as well as recommendations are also pointed out. This book aims to provide fundamental guidelines for researchers and engineers on how to integrate and apply SC, ML, DL, and OA techniques in the domain of geoengineering and geoscience.

References

Darabi A, Ahangari K, Noorzad A, Arab A (2012) Subsidence estimation utilizing various approaches - A case study: Tehran No. 3 subway line. Tunn Undergr Sp Technol 31:117–127. https://doi.org/10.1016/j.tust.2012.04.012

Dikshit A, Pradhan B, Alamri AM (2020) Pathways and challenges of the application of artificial intelligence to geohazards modelling. Gondwana Res

Gupta R, Tanwar S, Tyagi S, Kumar N (2020) Machine learning models for secure data analytics: A taxonomy and threat model. Comput Commun 153:406–440

LeCun Y, Bengio Y, Hinton G (2015) Deep learning. Nature 521:436–444. https://doi.org/10.1038/nature14539

Li Z, Zhang B, Wang Y et al (2014) Water pipe condition assessment: A hierarchical beta process approach for sparse incident data. Mach Learn 95:11–26. https://doi.org/10.1007/s10994-013-5386-z

Liang WJ, Zhuang DF, Jiang D, et al (2012) Assessment of debris flow hazards using a Bayesian Network. Geomorphology 171–172:94–100. https://doi.org/10.1016/j.geomorph.2012.05.008

Liu C, Yu Z, Zhao S (2021) A coupled SPH-DEM-FEM model for fluid-particle-structure interaction and a case study of Wenjia gully debris flow impact estimation. Landslides 18:2403–2425. https://doi.org/10.1007/s10346-021-01640-6

Luo HY, Shen P, Zhang LM (2019) How does a cluster of buildings affect landslide mobility: a case study of the Shenzhen landslide. Landslides 16:2421–2431. https://doi.org/10.1007/s10346-019-01239-y

Meng W, Li W, Zhou J (2021) Enhancing the security of blockchain-based software defined networking through trust-based traffic fusion and filtration. Inf Fusion 70:60–71. https://doi.org/10.1016/j.inffus.2020.12.006

Mohammadi SD, Naseri F, Alipoor S (2015) Development of artificial neural networks and multiple regression models for the NATM tunnelling-induced settlement in Niayesh subway tunnel, Tehran. Bull Eng Geol Environ 74:827–843. https://doi.org/10.1007/s10064-014-0660-2

Neaupane KM, Adhikari NR (2006) Prediction of tunneling-induced ground movement with the multi-layer perceptron. Tunn Undergr Sp Technol 21:151–159. https://doi.org/10.1016/j.tust.2005.07.001

Reichstein M, Camps-Valls G, Stevens B et al (2019) Deep learning and process understanding for data-driven Earth system science. Nature 566:195–204. https://doi.org/10.1038/s41586-019-0912-1

Sheil BB, Suryasentana SK, Mooney MA, Zhu H (2020) Machine learning to inform tunnelling operations: Recent advances and future trends. Proc Inst Civ Eng—Smart Infrastruct Constr 1–18. https://doi.org/10.1680/jsmic.20.00011

Shi S, Zhao R, Li S et al (2019) Intelligent prediction of surrounding rock deformation of shallow buried highway tunnel and its engineering application. Tunn Undergr Sp Technol 90:1–11. https://doi.org/10.1016/j.tust.2019.04.013

Shreyas SK, Dey A (2019) Application of soft computing techniques in tunnelling and underground excavations: state of the art and future prospects. Innov Infrastruct Solut 4:1–15. https://doi.org/10.1007/s41062-019-0234-z

Solomatine DP, Xue Y (2004) M5 Model trees and neural networks: Application to flood forecasting in the upper reach of the Huai River in China. J Hydrol Eng 9:491–501. https://doi.org/10.1061/(asce)1084-0699(2004)9:6(491)

Yi YL, Cao P, Pu CZ (2010) Multi-factorial comprehensive estimation for jinchans deep typical rockburst tendency. Sci. Technol. Rev. 28:76–80

Zhan LT, Guo XG, Sun QQ, et al (2021) The 2015 Shenzhen catastrophic landslide in a construction waste dump: analyses of undrained strength and slope stability. Acta Geotech 16:1247–1263. https://doi.org/10.1007/s11440-020-01083-8

Zhang W, Han L, Gu X, et al (2020a) Tunneling and deep excavations in spatially variable soil and rock masses: A short review. Undergr Sp

Zhang W, Tang L, Li H et al (2020) Probabilistic stability analysis of Bazimen landslide with monitored rainfall data and water level fluctuations in Three Gorges Reservoir, China. Front Struct Civ Eng 14:1247–1261. https://doi.org/10.1007/s11709-020-0655-y

Zhang W, Zhang R, Fu Y, et al (2018) 2D and 3D numerical analysis on strut responses due to one-strut failure. Geomech Eng 15:965–972. https://doi.org/10.12989/gae.2018.15.4.965

Chapter 2
Soft Computing

As indicated in the introduction part and the Fig. 2.1 plot, definition of SC has some overlaps with ML, DL, and OA. It uses component fields of study in: Fuzzy logic, Machine learning, Probabilistic reasoning, Evolutionary computation, Perceptron, Genetic algorithms, Differential algorithms, Support vector machines, Metaheuristics, Swarm intelligence, Ant colony optimization, Particle optimization, Bayesian networks, Artificial neural networks, Expert systems, etc. As a field of mathematical and computer study, SC has been around since the 1990s. The inspiration was the human mind's ability to form real-world solutions to problems through approximation. It contrasts with possibility, an approach that is used when there is not enough information available to solve a problem. In contrast, SC is used where the problem is not adequately specified for the use of conventional math and computer techniques. It has numerous real-world applications in domestic, commercial and industrial situations.

2.1 What is Soft Computing

Prior to 1994 when Zadeh (1994) firstly defined "soft computing ", the currently-handled concepts used to be referred to in an isolated way, whereby each was spoken of individually with an indication of the use of fuzzy methodologies. Although the idea of establishing the area of soft computing dates back to 1990 (Zadeh 2001), it was in Zadeh (1994) that Zadeh established the definition of soft computing in the following terms:

"Basically, soft computing is not a homogeneous body of concepts and techniques. Rather, it is a partnership of distinct methods that in one way or another conform to its guiding principle. At this juncture, the dominant aim of soft computing is to exploit the tolerance for imprecision and uncertainty to achieve tractability, robustness and low solutions cost. The principal constituents of soft computing are fuzzy logic, neurocomputing, and probabilistic reasoning, with the latter subsuming genetic

W. Zhang et al., *Application of Soft Computing, Machine Learning, Deep Learning and Optimizations in Geoengineering and Geoscience*,
https://doi.org/10.1007/978-981-16-6835-7_2

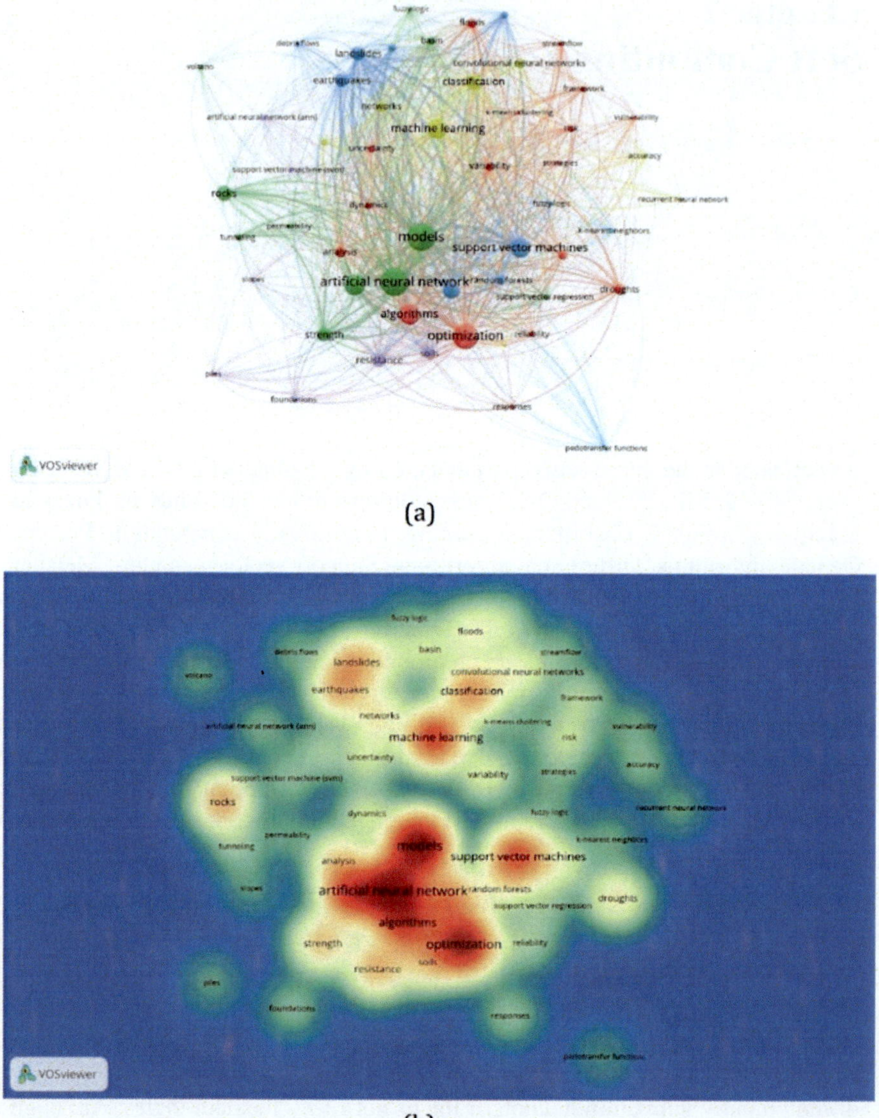

Fig. 2.1 VOSviewer plots for main SC methods: **a** network visualization **b** density visualization

algorithms, belief networks, chaotic systems, and parts of learning theory. In the partnership of fuzzy logic, neurocomputing, and probabilistic reasoning, fuzzy logic is mainly concerned with imprecision and approximate reasoning; neurocomputing with learning and curve-fitting; and probabilistic reasoning with uncertainty and belief propagation".

Therefore, it is clear that rather than a precise definition for soft computing, it is instead defined by extension, by means of different concepts and techniques which attempt to overcome the difficulties which arise in real problems which occur in a world which is imprecise, uncertain and difficult to categorize.

There have been various subsequent attempts to further hone this definition, with differing results, and among the possible alternative definitions, perhaps the most suitable is the one presented in Li et al. (1998): "Every computing process that purposely includes imprecision into the calculation on one or more levels and allows this imprecision either to change (decrease) the granularity of the problem, or to "soften" the goal of optimalisation at some stage, is defined as to belonging to the field of soft computing".

Soft computing is considered to be the antithesis of what we might call hard computing. It could therefore be seen as a series of techniques and methods so that real practical situations could be dealt with in the same way as humans deal with them, i.e. on the basis of intelligence, common sense, consideration of analogies, approaches, etc. In this sense, soft computing is a family of problem-resolution methods headed by approximate reasoning and functional as well as optimization approximation methods, including search methods. Soft computing is therefore the theoretical basis for the area of intelligent systems and it is evident that the difference between the area of artificial intelligence and that of intelligent systems is that the first is based on hard computing and the second on soft computing.

2.2 Components of Soft Computing

From this other viewpoint on a second level, soft computing can be then expanded into other components which contribute to a definition by extension, such as the one given in the last section. From the beginning (Bonissone 2002), the components considered to be the most important in this second level are probabilistic reasoning, fuzzy logic and fuzzy sets, neural networks, and genetic algorithms (GA), which because of their interdisciplinary, applications and results immediately stood out over other methodologies such as the previously mentioned chaos theory, evidence theory, etc. The popularity of GA, together with their proven efficiency in a wide variety of areas and applications, their attempt to imitate natural creatures (e.g. plants, animals, humans) which are clearly soft (i.e. flexible, adaptable, creative, intelligent, etc.), and especially the extensions and different versions, transform this fourth second-level ingredient into the well-known evolutionary algorithms (EA) which consequently comprise the fourth fundamental component of soft computing, as shown in the following diagram:

(1) From the first level and beginning with approximate reasoning methods, when we only concentrate on probabilistic models, we encounter the Dempster-Shafer theory and Bayesian networks. However, when we consider probabilistic methods combined with fuzzy logic, and even with some other multi-valued logics, we encounter what we could call hybrid probabilistic models, fundamentally probability theory models for fuzzy events, fuzzy event belief models, and fuzzy influence diagrams.

(2) When we look at the developments directly associated with fuzzy logic, fuzzy systems and in particular fuzzy controllers stand out. Then, arising from the combination of fuzzy logic with neural networks and EA are fuzzy logic-based hybrid systems, the foremost exponents of which are fuzzy neural systems, controllers adjusted by neural networks (neural fuzzy systems which differ from the previously mentioned fuzzy neural systems), and fuzzy logic-based controllers which are created and adjusted with EA.

(3) Moving through the first level to the other large area covered by soft computing (functional approach/optimization methods) the first component which appears is that of neural networks and their different models. Arising from the interaction with fuzzy logic methodologies and EA methodologies are hybrid neural systems, and in particular fuzzy control of network parameters, and the formal generation and weight generation in neural networks.

(4) The fourth typical component of soft computing and perhaps the newest yet possibly most up-to-date is that of EA, and associated with these are four large, important areas: evolutionary strategies, evolutionary programming, GA, and genetic programming. If we were only to focus on these last areas, we could consider that in this case the amalgam of methodologies and techniques associated with soft computing culminate in three important lines: fuzzy genetic systems, bioinspired systems, and applications for the fuzzy control of evolutionary parameters.

On further examination of this last component some additional considerations are needed. Firstly, independently of the broad-minded approach adopted to contemplate what can be embraced by fuzzy genetic systems, bioinspired systems, and fuzzy control applications on evolutionary parameters, other important topics are missing from this description. Secondly, if we are referring in particular to bioinspired systems, it is clear that not only are they the product of fuzzy logic, neural networks or EA (with all the variants that we can consider for these three components) but also that other extremely important methodologies are involved in them.

Figure 2.1 provides the VOSviewer plots for main SC methods. Figure 2.1a is the network visualization, in which the keyword co-occurrence network of SC methods was constructed by the VOSviewer software, which is a software tool for constructing and visualizing bibliometric networks. These networks may for instance include journals, researchers, or individual publications, and they can be constructed based on citation, bibliographic coupling, co-citation, or co-authorship relations. It also offers text mining functionality that can be used to construct and visualize co-occurrence networks of important terms extracted from a body of scientific literature (https://

www.vosviewer.com/). The size of the nodes and words represents the weights of the nodes. The bigger the node and word are, the larger the weight is. The distance between two nodes reflects the strength of the relation between two nodes. A shorter distance generally reveals a stronger relationship. The line between two keywords represents that they have appeared together. The thicker the line is, the more co-occurrence they have. The nodes with the same color belong to a cluster. VOSviewer divided the keywords of SC-related publications into 6 clusters, i.e., green cluster ball for ANN, red cluster ball for OA, dark blue cluster ball for SVR, purple cluster ball for SVM, yellow cluster ball for CNN and RNN, and light blue cluster ball for k-nearest neighbor. The link strength between two nodes refers to the frequency of co-occurrence. It can be used as a quantitative index to depict the relationship between two nodes. The total link strength of a node is the sum of link strengths of this node over all the other nodes. The node, "ANN", has thicker lines with "models", "optimization", "algorithm", "machine learning", "SVM", "classification", "rocks", "earthquakes", "landslides" and "droughts".

Figure 2.1b is the density visualization. Each node in the keywords density visualization plat has a color that relies on the density of items at that node. In other words, the color of a node depends on the number of items in the neighborhood of the node. The keywords in red color area appear more frequently; on the contrary, the keywords in green color area appear less frequently. Density views are especially useful for understanding the overall structure of a map and drawing attention to the most important areas in the map. From Fig. 2.1b, we can see the research focuses of SC study intuitively. "ANN", "models", "algorithms", "optimization", "SVM", "machine learning" turn out to be important. These keywords are the core keywords in the SC study.

2.3 Heuristics and Metaheuristics

As stated in Jiménez et al. (2003), since the fuzzy boom of the 1990s, methodologies based on fuzzy sets (i.e. soft computing) have become a permanent part of all areas of research, development and innovation, and their application has been extended to all areas of our daily life: health, banking, home, and are also the object of study on different educational levels. Similarly, there is no doubt that thanks to the technological potential that we currently have, computers can handle problems of tremendous complexity (both in comprehension and dimension) in a wide variety of new fields.

As we mentioned above, since the mid 1990s, GA (or EA from a general point of view) have proved to be extremely valuable for finding good solutions to specific problems in these fields, and thanks to their scientific attractiveness, the diversity of their applications and the considerable efficiency of their solutions in intelligent systems, they have been incorporated into the second level of soft computing components.

EA, however, are merely another class of heuristics, or metaheuristics, in the same way as Tabu Search, Simulated Annealing, Hill Climbing, Variable Neighbourhood Search, Estimation Distribution Algorithms (EDA), Scatter Search, GRASP, Reactive Search and very many others are. Generally speaking, all these heuristic algorithms (metaheuristics) usually provide solutions which are not ideal, but which largely satisfy the decision-maker or the user. When these act on the basis that satisfaction is better than optimization, they perfectly illustrate Zadeh's famous sentence: "...in contrast to traditional hard computing, soft computing exploits the tolerance for imprecision, uncertainty, and partial truth to achieve tractability, robustness, low solution-cost, and better rapport with reality".

Consequently, among the soft computing components, instead of EA (which can represent only one part of the search and optimization methods used), heuristic algorithms and even metaheuristics should be considered.

There is usually controversy about the difference between metaheuristics and heuristics, and while it is not our intention here to enter into this debate, we are interested in offering a brief reflection on both concepts. The term heuristics comes from the Greek word "heuriskein", the meaning of which is related to the concept of finding something.

It is therefore clear that metaheuristics are more broad-brush than heuristics. In the sections which follow, we will focus on the concept of metaheuristics, and will start by pointing out that in the terms that we have defined, certain metaheuristics will always be better than others in terms of their performance when it comes to solving problems.

In order to achieve the best performance of the metaheuristics, it is desirable for them to have a series of "good properties" which include simplicity, independence, coherence, effectiveness, efficiency, adaptability, robustness, interactivity, diversity, and autonomy (Melian et al. 2003). In view of their definition and the series of desirable characteristics, it is both logical and obvious that EA are to be found among metaheuristics and they are therefore well placed with the other second-level soft computing components to facilitate the appearance of new theoretical and practical methodologies, outlines, and frameworks for a better understanding and handling of generalized imprecision in the real world.

2.4 Hybrid Metaheuristics in Soft Computing

In this section, we will consider the three main previously mentioned groups of metaheuristics. From these, we will then describe the new metaheuristics which have emerged, briefly dwelling on the less developed or less popular ones because they are more recent.

1. Evolutionary Metaheuristics. These metaheuristics are by far the most popular and define mechanisms for developing an evolution in the search space of the sets of solutions in order to come close to the ideal solution with elements which

will survive in successive generations of populations. In the context of soft computing, the hybridizations which take these metaheuristics as a reference are fundamental.

2. Relaxation metaheuristics. A real problem may be relaxed when it is simplified by eliminating, weakening or modifying one of its characteristic elements. Relaxation metaheuristics are strategies for relaxing the problem in heuristic design, and which are able to find solutions for problems which would otherwise have been very difficult to solve without the use of this methodology. Examples of these are rounding up or down or adjustments in nature, as occurs when an imprecisely and linguistically-expressed quantity is associated to an exact numerical value. From this point of view, a real alternative is to flexibilize exact algorithms, introducing fuzzy stop criteria, which eventually leads to rule-based relaxation metaheuristics; admitting the vagueness of coefficients, justifying algorithms for resolving problems with fuzzy parameters, and relaxing the verification of restrictions, allowing certain violations in their fulfillment.
3. Search metaheuristics. Generally speaking, these are probably the most important metaheuristics, and their basic operation consists in establishing strategies for exploring the solution space of the problem and iterating the starting-point solutions. Although at first sight they might appear to be similar to evolutionary searches, they are not since evolutionary searches base their operation on the evolution of a population of individuals in the search space. These metaheuristics are usually described by means of various metaphors, which classify them as bioinspired, sociological, based on nature, etc. and this makes them extremely popular.

If, however, the search procedure is performed using various metaheuristics, there is always the possibility of cooperation between these (Cruz 2005), and therefore the generalization of everything described so far to the context of parallelism, something which is obviously beyond the sphere of this paper but which it is interesting to reflect on since with the proliferation of parallel computing, more powerful work stations and faster communication networks, parallel implementations of metaheuristics have emerged as something natural and provide an interesting alternative for increasing the speed of the search for solutions. Various strategies have correspondingly been proposed and applied and these have proved to be very efficient for resolving large-scale problems and for finding better solutions than those of their sequential counterparts due to the division of the search space, or because they have improved the intensification and diversification of the search. As a result, parallelism (and therefore multiple metaheuristics) not only constitutes a way of reducing the execution times of individual metaheuristics, but also of improving their effectiveness and robustness.

2.5 A Data-Driven Nonparametric Explainable MARS Model

Multivariate Adaptive Regression Splines (MARS) was first proposed by Friedman (2007) as an adaptive procedure for regression, which works well in high-dimensional analysis. It is a non-parametric statistical method based on a divide and conquer strategy in which the training data sets are partitioned into separate piece-wise linear segments (splines) of differing gradients (slope). In general, the splines are connected smoothly together by weighted summation to handle linear structures, in some special cases, these piece-wise curves, also known as basis functions (BFs) can be involved into multiplications as well to obtain higher-order terms fitting nonlinear behaviors.

The form of the piece-wise linear basis functions are shown below:

$$(x-t)_+ = \begin{cases} x-t, if x > t, \\ 0, otherwise, \end{cases}$$

And (2.1)

$$(t-x)_+ = \begin{cases} t-x, if x < t, \\ 0, otherwise, \end{cases}$$

where the subscript "+" means positive part.

As shown in Fig. 2.2, each one of the basis functions (BFs) $(x-1)_+$ and $(1-x)_+$ is piece-wise linear, the interface point between the two pieces is called knot, in this case, the knot value is 1. The two functions together are called as a reflected pair in the discussion below. As the idea is to form reflected pairs for each input X_j with knots at each observed value X_{ij} of that input, the collection of basis functions is

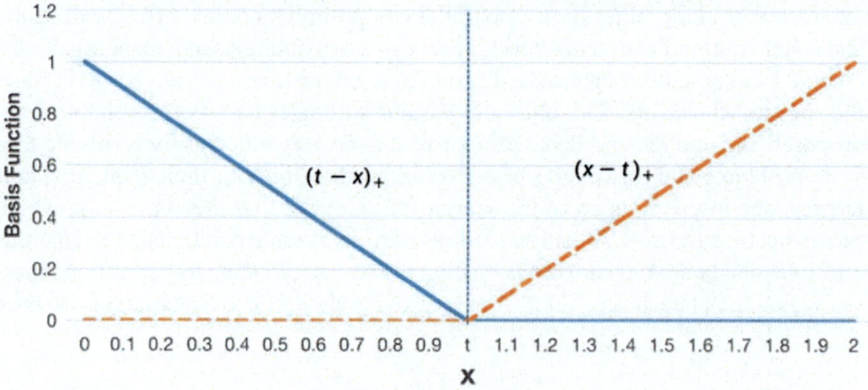

Fig. 2.2 The example of basis functions used by MARS

$$C = \left\{ (X_j - t)_+, (t - X_j)_+ \right\}_{j=1,2,3,4,\ldots,N}^{t \in \{x_{1j}, x_{2j}, x_{3j}, \ldots, x_{Nj}\}} \tag{2.2}$$

Forward stepwise linear regression is applied in the model-building process, but instead of directly using the original inputs X, the functions from the set C and the their interactions are used. Therefore, the form of the model is.

$$F(X) = \beta_0 + \sum_{m=1}^{M} \beta_m h_m(X) \tag{2.3}$$

where each $h_m(X)$ is a function in the set C, or the products of two or more such functions. The model is started with a constant function $h_0 = 1$. In each step of the forward phase, our goal is to obtain a new basis function pair from all products of a function h_m in the current model K with one of the reflected pairs in the set C to produce the largest reduction in the training error. The winning product of each step is added to the model until reaches its preset maximum number of terms. Considering a current model with M basis functions, the next pair to be added to the model is in the form of

$$\hat{\beta}_{M+1} h_l(X) * (X_j - t)_+ + \hat{\beta}_{M+2} h_l(X) * (t - X_j)_+, \; h_l \in K, \tag{2.4}$$

with each β being estimated by minimizing the residual sum-of-squares, that is, by standard linear regression.

Figure 2.3 shows an example illustrating how MARS algorithm would make use of basis functions to fit provided data patterns. The MARS mathematical equation is

$$y = -5.0875 - 2.7678h_1(X) + 0.5540h_2(X) + 1.1900h_3(X) \tag{2.5}$$

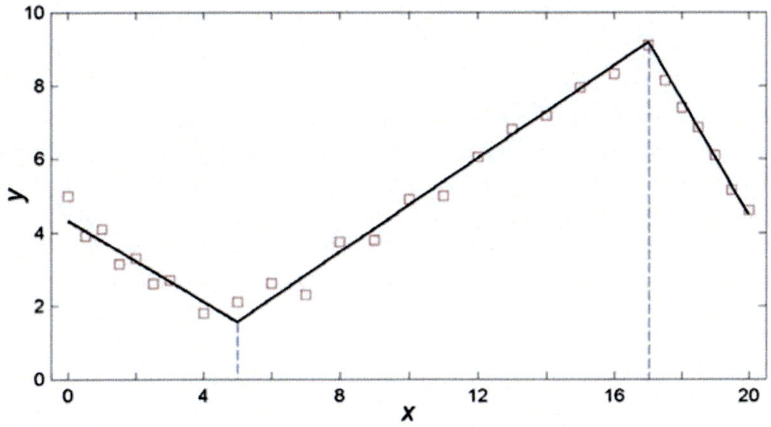

Fig. 2.3 Knots and basis functions for a simple MARS example

in which $h_1(X) = (x - 17)_+$, $h_2(X) = (17 - x)_+$, $h_3(X)(x - 5)_+$. The knots are located at x = 5 and 17, which cut the x range into three intervals where different linear relationships are identified.

The forward modeling strategy in MARS is hierarchical, as a high-order interaction will only likely exist if some of its lower order "footprints" already exist. For example, a three-way product can only be added into the model when one of its two-way products exists in the model. This strategy helps to avoid the search over an exponentially growing space of alternatives.

Generally, the upper limit on the order of interactions is set to be two in the MARS procedure to maintain its high interpretation, but not produce too much negative impact on its adaptiveness on fitting linear structures and non-linear behaviors. When the reflected pair in the set C is multiplied by the constant function $h_0 = 1$, a linear term is obtained, otherwise the nonlinear structure is formed. Each input is restricted to appear at most once in a term to prevent the sharply increase or decrease near the boundaries of the feature space.

There is no need of presetting the underlying functional relationships between dependent and independent variables, instead, MARS allows the model itself to determine the best suited linear and non-linear structures by adding one knot and its corresponding pair of BFs to give the maximum reduction in sum-of-squares residual error in each step. This process ends at the predetermined situation and usually leads to a purposely overfitting model.

Therefore, a backward deletion procedure is applied, which removes the term that contribute least on reducing the training error at each step, obtaining an estimated best model $\widehat{f_\lambda}$ of each size λ. Generalized cross-validation is used to optimize λ, the criterion is defined as

$$\mathrm{GCV}(\lambda) = \frac{\sum_{i=1}^{N} (y_i - \widehat{f_\lambda}(x_i))^2}{(1 - M(\lambda)/N)^2}, \tag{2.6}$$

$$M(\lambda) = r + cK, \tag{2.7}$$

where M(λ) represents the effective number of parameters in the model, r is the number of terms in the model, K is the number of knots that were selected in the forward process, and c is the number of parameters used for determining one optimal knot value. Some mathematical and simulation results suggest that three parameters should be used for selecting a knot in a piecewise linear regression, so c = 3 in general. But in the case of the model being restricted to be additive, c = 2 is used.

Because of its high flexibility, MARS is usually viewed as the generalization of step-wise linear regression. But other than that, it can also be treated as a modification of the CART method.

Although they might seem quite different, MARS can actually be turned into CART by applying the changes below.

- Rewrite the piecewise linear basis functions (Eq. 2.1) as step functions

$$\begin{aligned} &I(x-t>0) \\ &\text{And} \\ &I(x-t\leq 0). \end{aligned} \tag{2.8}$$

- Replace the model term by the interaction when it is involved in a multiplication by a candidate term, so it is not available for further interactions.

The first change results in a step function multiplied by a pair of reflected step functions in each step, which is equivalent to splitting a node at the step. The second one corresponds to the fact that the same node can only be split once in CART model, which causes CART not have the ability to handle additive structures. By allowing further interactions of a certain model term even when it is involved in a multiplication by a candidate term, MARS forgoes the binary-tree representation of the CART model, thus gains the ability to fit addictive structures in a much more effective way.

Similar to CART, MARS not only works very well in regression, but can also handle classification problems efficiently.

Generally, for binary classification problems, one can encode the output as 0/1 and then treat it as a regression problem. For the multi-classification problems, the indicator response approach can be used. The K response classes are coded via 0/1 indicator variables, and then a multi-response MARS regression is performed, in which a common set of basis functions are used for all response variables. Finally, the output is the class with the largest predicted value. However, potential masking problems exist in this approach, the better way is to use "optimal scoring" method (Hastie et al. 2017).

Specifically, there is a hybrid of MARS called PolyMARS designed by Kooperberg et al. (1997) to handle classification problems. It uses the multiple logistic framework, by which it grows the model in a forward stagewise fashion like MARS, but at each stage, the quadratic approximation of the polynomial log-likelihood is used to search for the next basis function pair. Once found, the enlarged model is fit by maximum likelihood, and the process is repeated until reach preset goal is reached (Hastie et al. 2017).

As for the MARS use, Fig. 2.4 provides the VOSviewer plots for network visualization and density visualization, in which the explanation about the network and density visualization is the same. For brevity, these interpretations are ignored. Figure 2.4c provided the overlay visualization The overlay visualization is identical to the network visualization except that items are colored differently via use of the color bar. Herein the color bar in the lower right corner of the visualization represents the years of publication, with correspondence between colors.

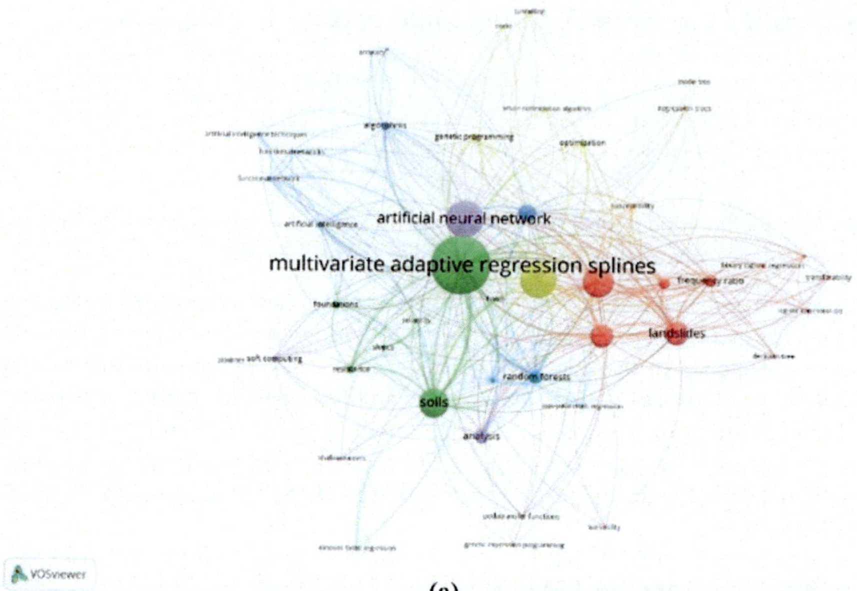

(a)

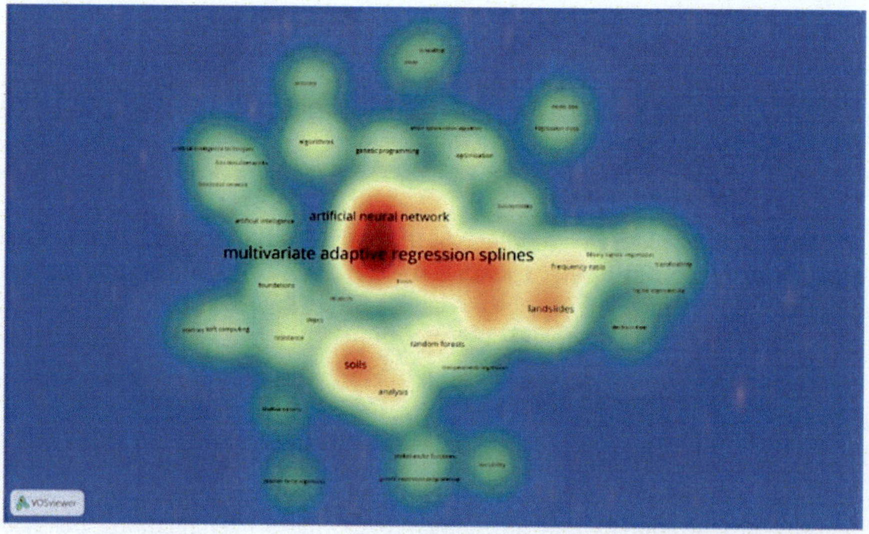

(b)

Fig. 2.4 VOSviewer plots for MARS method: **a** network visualization **b** density visualization **c** overlay visualization

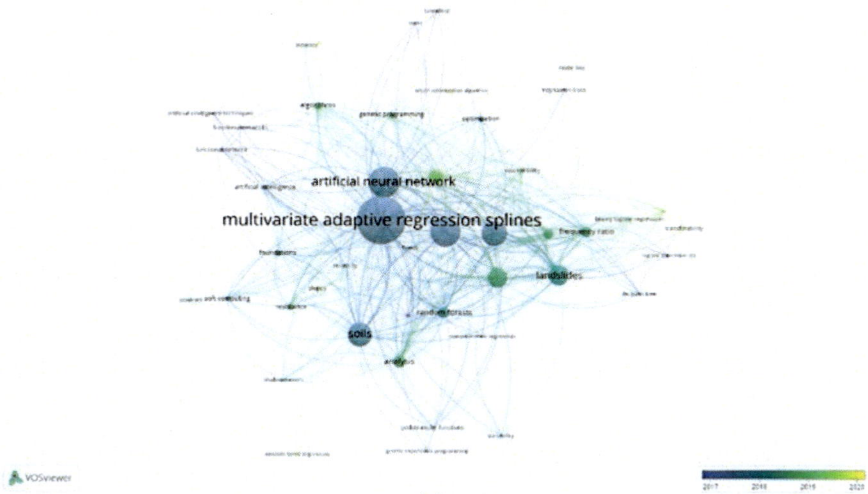

(c)

Fig. 2.4 (continued)

References

Bonissone P (2002) Hybrid soft computing for classification and prediction applications: Keynote address. Lecture Notes in Computer Science (including subseries Lecture Notes in Artificial Intelligence and Lecture Notes in Bioinformatics). Springer, Berlin, Heidelberg, pp 352–353

Cruz CA (2005) Estrategias coordinadas paralelas basadas en soft-computing para la solución de problemas de optimización

Friedman JH (2007) Multivariate adaptive regression splines. Ann Stat 19:1–67. https://doi.org/10.1214/aos/1176347963

Hastie T, Tibshirani R, Tibshirani RJ (2017) Extended comparisons of best subset selection, forward stepwise selection, and the Lasso

Jiménez F, Gómez Skarmeta AF, Sánchez G, Cadenas JM (2003) Fuzzy sets based heuristics for optimization: Multi-objective evolutionary fuzzy modeling. In: Verdegay J-L (ed) Springer. Berlin Heidelberg, Berlin, Heidelberg, pp 221–234

Kooperberg C, Bose S, Stone CJ (1997) Polychotomous regression. J Am Stat Assoc 92:117–127. https://doi.org/10.1080/01621459.1997.10473608

Li X, Ruan D, Van Der Wal AJ (1998) Discussion on soft computing at FLINS'96. Int J Intell Syst 13:287–300. https://doi.org/10.1002/(sici)1098-111x(199802/03)13:2/3%3c287::aid-int10%3e3.0.co;2-4

Melian B, Moreno-Pérez J, Moreno-Vega J (2003) Metaheuristicas: Una visión global. Intel Artif Rev Iberoam Intel Artif ISSN 1137–3601, null 7, Nº 19, 2003, pags 7–28 7:

Zadeh LA (1994) Soft computing and fuzzy logic. IEEE Softw 11:48–56. https://doi.org/10.1109/52.329401

Zadeh LA (2001) Applied soft computing—foreword. Appl Soft Comput 1:1–2. https://doi.org/10.1016/s1568-4946(01)00003-5

Chapter 3
Machine Learning and Applications

3.1 Supervised Learning

Supervised learning, is defined by its use of labeled datasets to train algorithms that to classify data or predict outcomes accurately. As input data is fed into the model, it adjusts its weights until the model has been fitted appropriately. This occurs as part of the cross validation process to ensure that the model avoids overfitting or underfitting. Some methods used in supervised learning include neural networks, naïve bayes, linear regression, logistic regression, random forest, support vector machine (SVM), and more.

3.1.1 ANN

Artificial neural network (ANN) is a mathematical model which imitates the structure and function of biological neural networks (Agatonovic-Kustrin and Beresford 2000). It uses a large number of artificial neuron connections to perform calculations. The network is formed by many "neurons" connected to each other. Each neuron represents a specific output function, also named activation function. The connection between the neurons in each layer represents a weighted value of the signal passing through the connection, called weight coefficient, which is equivalent to the memory proportion of ANN. The output of the network is determined by the connection rules of the network, ANN can be understood as an approximation of a certain algorithm or function in nature.

Figure 3.1 describes a simple neural network structure that includes an input layer, a hidden layer, and an output layer. The cells in the hidden layer are fully connected to the input layer, while the output layer is fully connected to the hidden layer. If such a network has more than one hidden layer, it can be called a multilayer neural network.

W. Zhang et al., *Application of Soft Computing, Machine Learning, Deep Learning and Optimizations in Geoengineering and Geoscience*,
https://doi.org/10.1007/978-981-16-6835-7_3

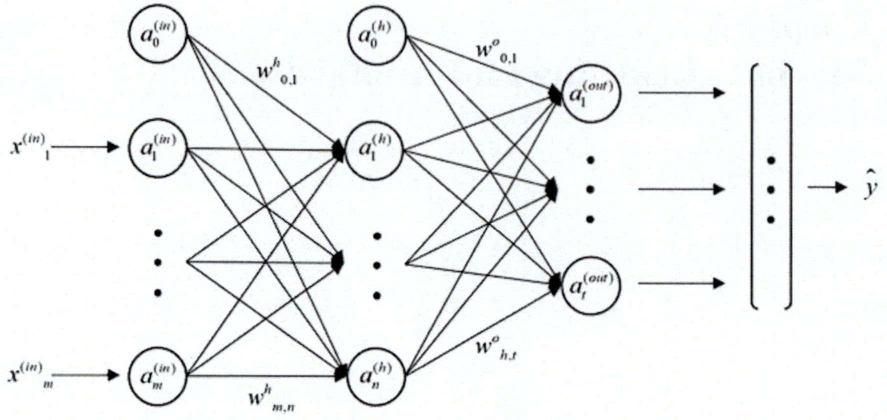

Fig. 3.1 Architecture of artificial neural network

As above, $a_i^{(l)}$ represents the *i* neuron in the *l* layer. Superscript "*in*" denotes the input layer, superscript *h* denotes the hidden layer and superscript "*out*" represents the output layer. The $a_0^{(in)}$ and $a_0^{(h)}$ are deviation value, defined as 1, and the input layer is the input value Plus the deviation value.

Figure 3.2 provides the VOSviewer plots for some main supervised learning methods, in which the keyword co-occurrence network of ANN was constructed by the VOSviewer software. The size of the nodes and words represents the weights of the nodes. The bigger the node and word are, the larger the weight is. The distance between two nodes reflects the strength of the relation between two nodes. A shorter distance generally reveals a stronger relationship. The line between two keywords represents that they have appeared together. The thicker the line is, the more co-occurrence they have. The nodes with the same color belong to a cluster. VOSviewer divided the keywords of ANN-related publications into 7 clusters.

3.1.2 ELM

Extreme learning machine is a new feedforward neural network training method proposed by Huang et al. (2006). This method randomly generates the connection weights of the input layer, the hidden layer and the threshold value of the hidden layer neurons. It only needs to set the number of hidden layer neurons in the training process to obtain the unique optimal solution. Assuming there are *N* different samples, among which $x_i = [x_{i1}, x_{i2}, \cdots, x_{in}]^T \in R^n$, $t_i = [t_{i1}, t_{i2}, \cdots, t_{im}]^T \in R^m$, the standard single hidden layer feedforward neural network with hidden layer neurons *M* and activation function $g(x)$ can be expressed as

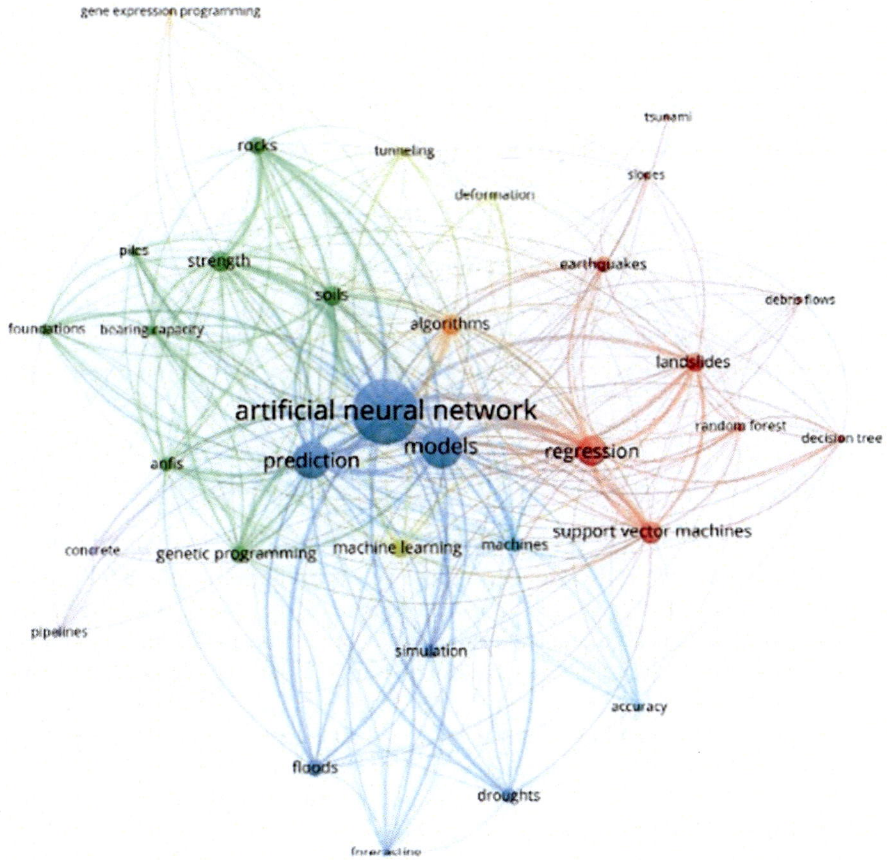

Fig. 3.2 Visualization of ANN algorithms

$$\sum_{i=1}^{M} \beta_i g(w_i \cdot x_i + b_i) = O_j, j = 1, \cdots, N, \tag{3.1}$$

where $w_i = [w_{i1}, w_{i2}, \cdots, w_{in}]^T$ represents the connection weight between the input layer node and the hidden layer node i;$\beta_i = [\beta_{i1}, \beta_{i2}, \cdots, \beta_{im}]^T$ represents the connection weight between the hidden layer node i and the output layer node; b_{i} represents the deviation coefficient of the hidden layer; $w_i \cdot x_j$ represents the inner product of w_i and x_j; $O_i = \left[O_{i1}, O_{i2}, \cdots, O_{jn}\right]^T$ represents the output values.

A standard single hidden layer feedforward neural network with hidden layer M and activation function $g(x)$ can approximate the training sample with zero error, when:

$$\sum_{i=1}^{M} \| o_j - t_j \| = 0 \tag{3.2}$$

Therefore, β_i, w_i, b_i meet the following formula:

$$\sum_{i=1}^{M} \beta_i g(w_i \cdot x_i + b_i) = y_i, j = 1, \ldots, N, \tag{3.3}$$

It can be expressed as $H\beta = Y$, where H is the output matrix of the neural network. Since the randomly selected input weights and hidden layer bias are fixed, network training is equivalent to solving the least squares solution $\widehat{\beta}$ of the following equation:

$$\min_{\beta} \| H\beta - Y \| \tag{3.4}$$

For the minimum value of the least square solution of the above linear system is $\widehat{\beta} = H^{+}Y$, where H^{+} is the Moore–Penrose generalized inverse of the hidden layer output matrix H.

The learning algorithm of the extreme learning machine mainly has the following three steps. Firstly, randomly set the connection weight$w$ of the input layer with the hidden layer and the bias b of hidden layer neurons. Secondly, calculate the output matrix H of the hidden layer. Lastly, calculate the weight of the output layer, $\widehat{\beta} = H^{+}Y$.

Figure 3.3 provides the Visualization of ELM algorithms, as well as the applications in geoengineering and geoscience.

3.1.3 DT

Decision trees are a way to represent rules underlying data with hierarchical, sequential structures that recursively partition the data (Murthy 1998). Decision trees are a common type of machine learning algorithm, which makes decisions based on a tree structure. A flow chart of establishing the decision tree is shown in Fig. 3.4. S represents the training sample set, A denotes the classification sample set, and N represents a classification leaf node.

Decision tree algorithms include Iterative Dichotomiser 3 (ID3), C4.5, Classification And Regression Tree (CART)ANN, Supervised Learning In Quest (SLIQ), etc. (Quinlan 1993; Breiman et al. 2017). Among these algorithms, the well-known and commonly used one is Quinlan's ID3 algorithm. C4.5 and CART are derived from the ID3 decision tree algorithm. The SLIQ algorithm is further obtained based on the modifications of the C4.5 decision tree classification algorithm. In C4.5 algorithm, the tree is constructed according to the depth-first strategy, while the structure

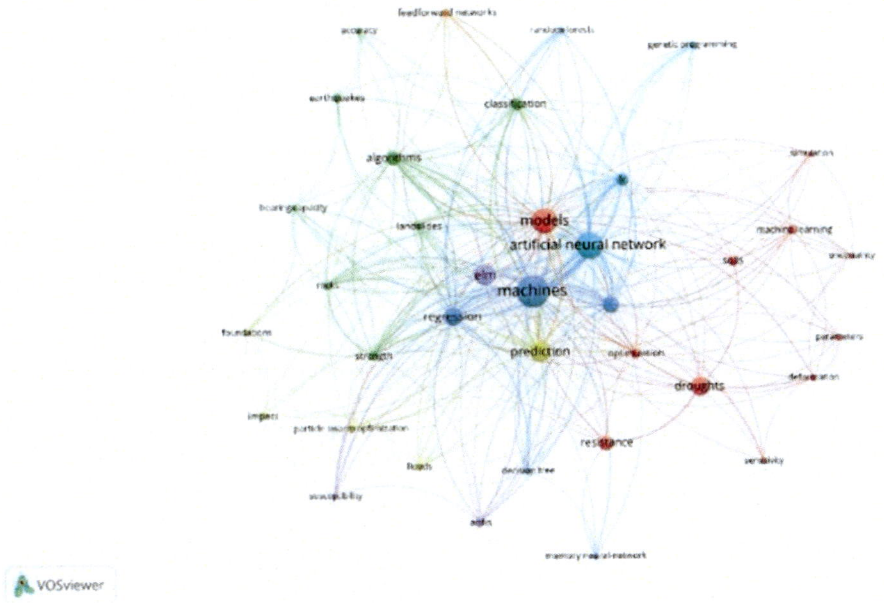

Fig. 3.3 Visualization of ELM algorithms

for SLIQ is constructed according to the breadth-first strategy (Niuniu 2010). The generation process of a decision tree has three main parts, namely feature selection, decision tree generation and pruning. The most critical issue is feature selection. Different splitting criteria have significant influence on the generalization error of the decision tree. Figure 3.5 provides the Visualization of DT algorithm.

3.1.4 LR

One of the oldest and simplest parametric statistical approaches is linear regression (Rashidi et al. 2019). This technique has been widely used in many studies (Wei et al. 2019; Elmousalami 2020; Ben Chaabene et al. 2020). Linear regression models are to find the best-fitted straight line that is also known as the "least squares regression line" (the best dotted line with the lowest error sum) between the independent variables and the dependent variables (Fig. 3.6). The advantage of this approach is its simplicity and transparency to find the linear relationship. However, this method may not be applicable for some complicated cases, where relationships between the independent variable (the cause or features) and the dependent variables are nonlinear. Figure 3.7 provides the Visualization of LR algorithm.

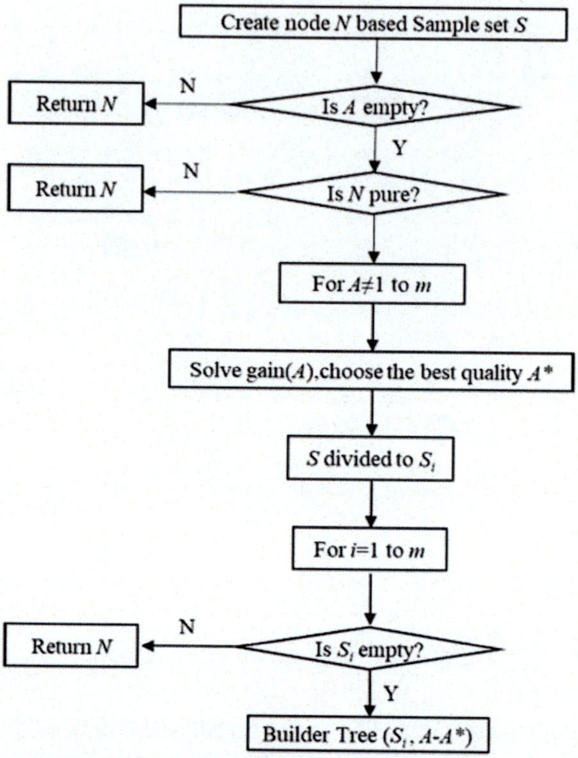

Fig. 3.4 Flow chart of decision tree development

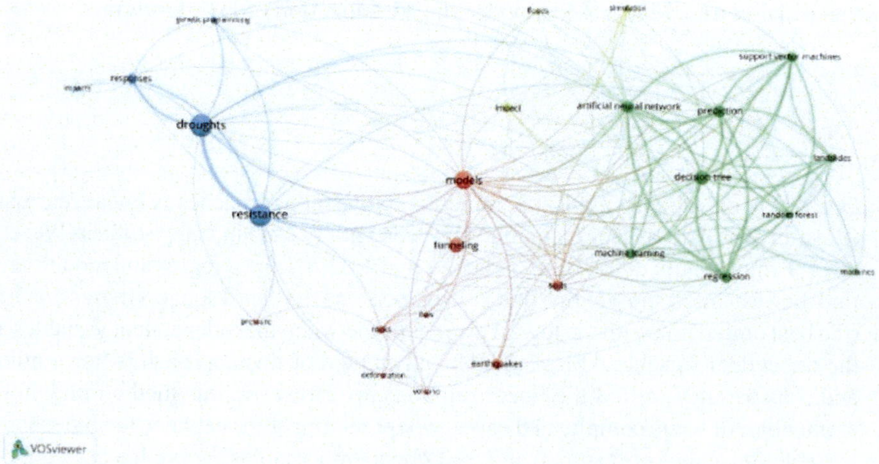

Fig. 3.5 Visualization of DT algorithm

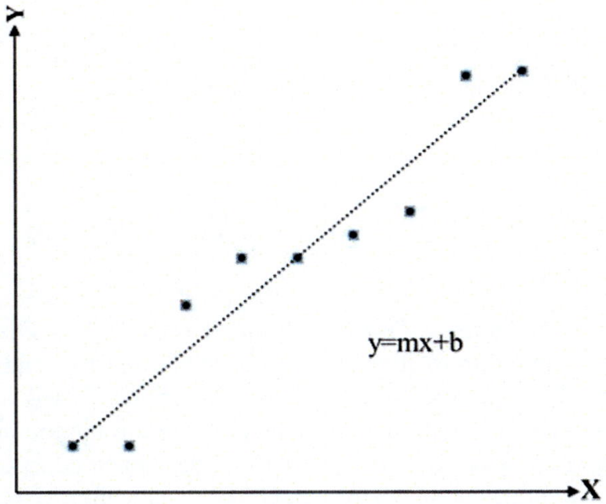

Fig. 3.6 Linear regression

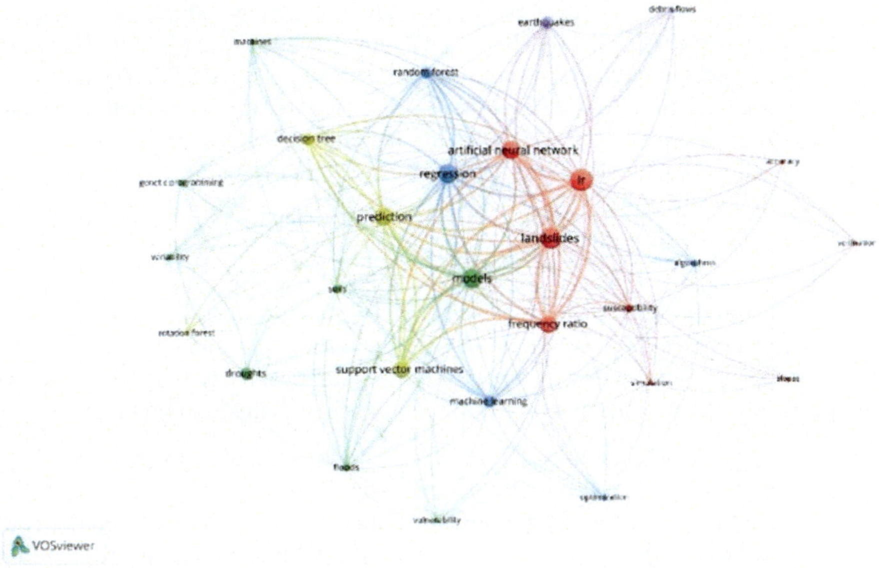

Fig. 3.7 Visualization of LR algorithm

3.1.5 GP

Genetic Programming (GP) is an automated method for creating computer programs starting from a high-level description of the problem to be solved (Oltean et al. 2009).

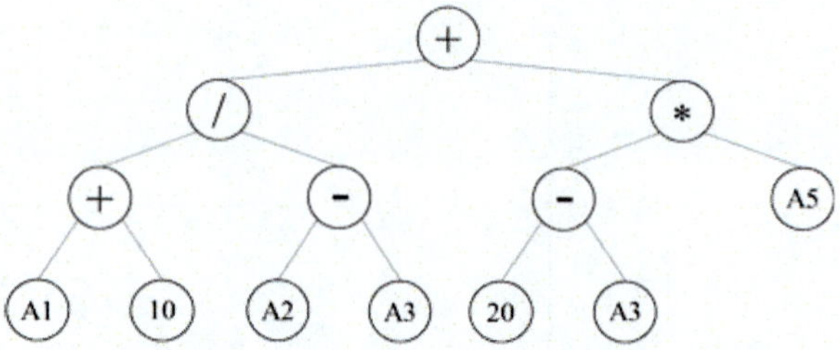

Fig. 3.8 Examples of standard genetic programming coding methods

(1) Genetic programming (Wieczorek and Czech 2000): The coding method mainly adopts a tree structure, with special crossover and mutation operators based on the tree structure, which is not only used for problem planning and discrete optimization, but also used for regression analysis and other fields. (Koza 1992; Angeline 1994) developed genetic programming to a great extent, and edited books on GP systematically. GP began to be widely used in automatic programming and numerical and combinatorial optimization problems.
(2) Coding and decoding: Genetic programming mainly uses syntax trees for coding instead of traditional linear coding. Figure 3.8 shows the tree coding of the expression $(A1 + 10) \div (A2 - A3) + (20 - A3) \times A5$. The expression contains variables $A1$,$A2$,$A3$,$A5$ and constants 10, 20. These variables and constants are located on the leaves of the tree and are called terminators. The symbols +,−, ×, ÷ are called operators. The terminators and operators together constitute the original set of genetic programming.

The advanced form of genetic programming can be constructed in a multi-component form.

In this case, the coding of GP is carried out by connecting multiple subtrees to a root node through specific rules as shown in Fig. 3.9. Each component is a subtree, and each subtree is a complete syntax tree. The form of the solution of each problem can be encoded into a tree structure through grammatical analysis in GP. At the same time, the decoding is deconstructed in accordance with the grammatical rules.

(3) Crossover and mutation: The basic operations of genetic programming include duplication, crossover, mutation and so on. Generally, the greater the fitness of an individual is, the greater the probability that it will be. The copy operation is to randomly select an individual based on fitness and copy it to the next-generation population. The crossover operation is to generate two offspring

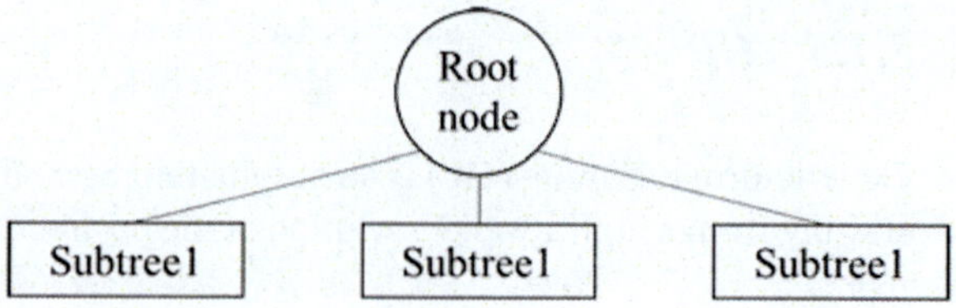

Fig. 3.9 Component-based genetic programming coding method

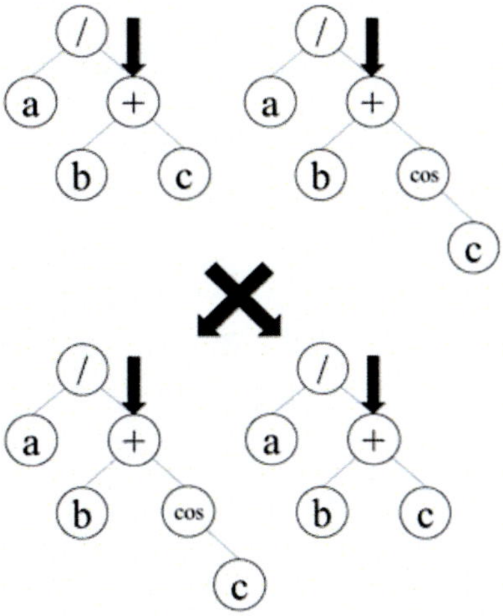

Fig. 3.10 Cross operation

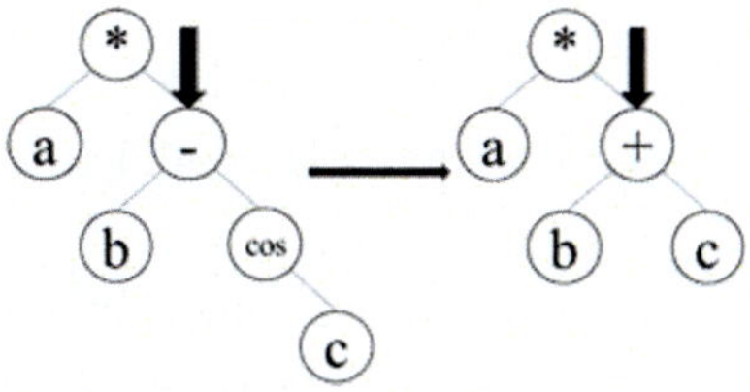

Fig. 3.11 Mutation operation

individuals from two randomly selected parent individuals. Figure 3.10 shows the process of two individuals performing crossover operations. Through mutation operation, a child individual is generated from a parent individual selected randomly based on fitness. As shown in Fig. 3.11, an individual has undergone a mutation operation.

(4) Fitness function and fitness: the concept of fitness which from nature is used in genetic programming to measure how likely each individual in the population is to help find the optimal solution in the optimization calculation. Individuals with higher fitness will have a higher probability of inheriting to the next generation, while individuals with lower fitness have a relatively small probability of inheriting to the next generation. The function that measures fitness is called fitness function.

(5) GP algorithm: The basic steps of genetic programming can be roughly divided into three steps: a. randomly generating the initial population; b. repeating the following sub-steps on the program population until the termination criterion is met. Each program individual is assigned a fitness degree. Three genetic operators (copy, crossover, mutation) are used to generate a new program population, and the individual to be processed is selected based on the probability value of

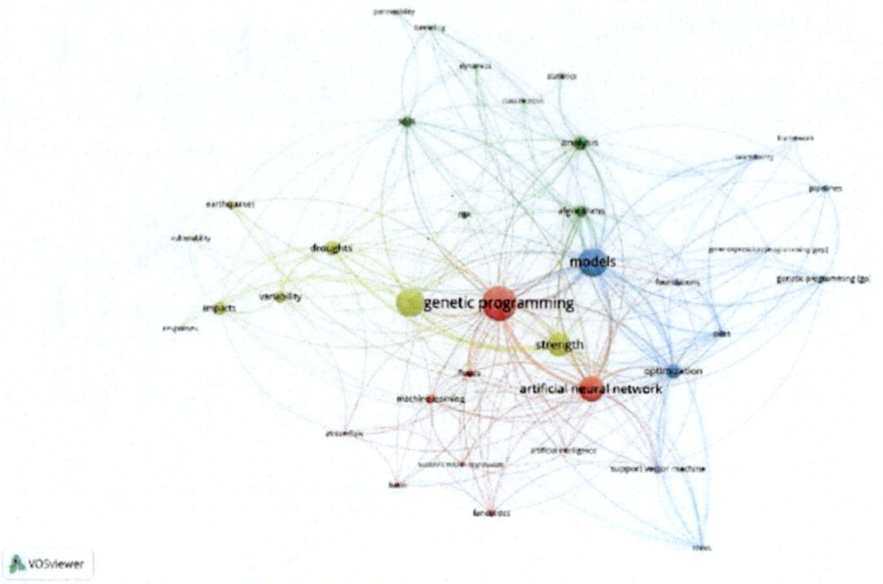

Fig. 3.12 Visualization of GP algorithm

fitness (reselection is allowed); c. returning the individual program determined by means of indicating the result as the running result of genetic programming.

Figure 3.12 provides the Visualization of GP algorithm.

3.1.6 NB

Naive Bayes (NB) is a classifier based on applying Bayes' theorem with absolute independence assumptions between attributes. As a result of the naive design and simplified assumptions, the NB model can work efficiently and it only requires a relatively small database for optimizing hyper-parameters (Bhargavi and Jyothi 2009). In general, the construction for a NB model contains five main steps: (i) collecting data, (ii) estimating a prior probability of each class, (iii) estimating means of classes, (iv) generating covariance matrices and finding the inverse and determinant for each class, (v) forming the discriminant function for each lass (Bhargavi and Jyothi 2009; Pham et al. 2017).

Given that $x = (x_1, x_2, ...x_n)$ is the input vector and $y = (y_1, y_2, ...y_K)$ is the output vector. The process for classification of NB model can be expressed as follows:

$$\hat{y} = \arg \max_{k=\{1,2,...,K\}} = \frac{P(y_k) \prod_{i=1}^{n} P(x_i|y_k)}{P(x)} \tag{3.5}$$

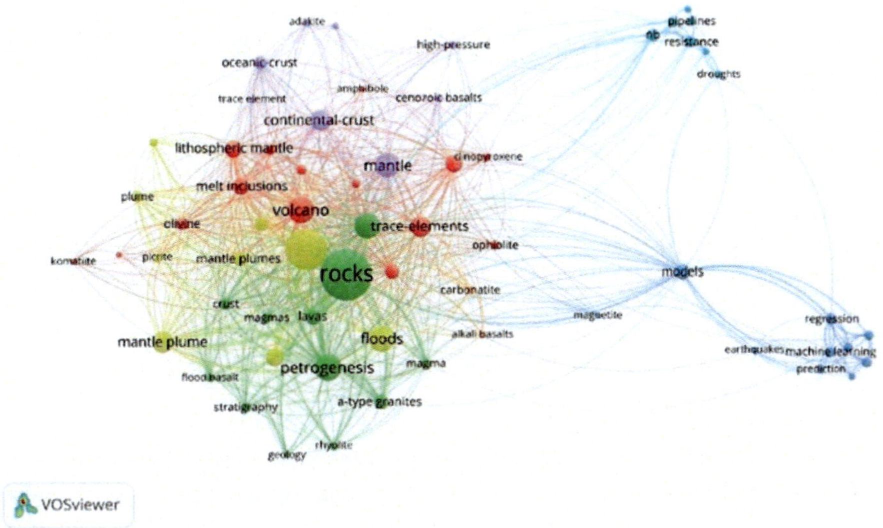

Fig. 3.13 Visualization of NB algorithm

where $\hat{y}$ is the class label; $P(y_k)$ is the prior probability of y_k; $P(x_i|y_k)$ is the conditional probability. Figure 3.13 provides the Visualization of NB algorithm.

3.1.7 SVM

Support vector machine (SVM) is one kind of the supervised learning algorithm, first proposed by Vapnik (2000) and gradually gains more and more popularity for its superior performance in dealing with classification and regression problems. In the following, some basic principles of the SVM model are presented.

Assume a specific training set containing l data points, such as (x_1, y_1), (x_2, y_2), …, (x_l, y_l), where $x_i \in R^n$ (i = 1,2,…, l) is the input variables and $y_i \in R$ (i = 1,2,…, l) is the corresponding target value. According to Xue et al. (2014), the aim for the support vector regression (SVR) is to solve an optimization problem as follows:

Minimize

$$\frac{1}{2}||w||^2 + C\sum_{i=1}^{l}(\xi_i + \xi_i^*) \tag{3.6}$$

Subjected to

$$\begin{cases} y_i - \sum_{j=1}^{n}\sum_{i=1}^{l} w_j x_{ji} - b \le \varepsilon + \xi_i \\ \sum_{j=1}^{n}\sum_{i=1}^{l} w_j x_{ji} + b - y_i \le \varepsilon + \xi_i^* \\ \xi_i, \xi_i^* \ge 0 \end{cases} \quad (3.7)$$

where

$$f(x) = \sum_{j=1}^{n} w_j x_j + b \text{ with } w \in R^n,\ \mathrm{b} \in \mathrm{R} \quad (3.8)$$

Some detailed information about these three equations is listed herein. n in Eq. (3.7) and Eq. (3.8) is the number of all variables for training and testing while b represents the bias. j is the weight for the l inputs to the corresponding outputs. The coefficient C guarantees both the simplicity and prediction accuracy of $f(x)$. When the errors between the predicted results and the observations are smaller than ε, this is within permission and no penalty is needed, where ε is named as the insensitive loss function shown in Fig. 3.14. Aside from ε, two slack variables, ξ_i and ξ_i^* to determine the degree of penalizing the samples, can be found in Fig. 3.14.

When constructing the SVR models, there are several common kernel functions, including linear kernel functions, polynomial kernel functions, radial basis functions (RBF), etc. (Zhang et al. 2016). To obtain an ideal SVR model, the available data is often divided into training set and the validation set, respectively. The training set is first utilized to build up a rough model and then the validation set is adopted to adjust

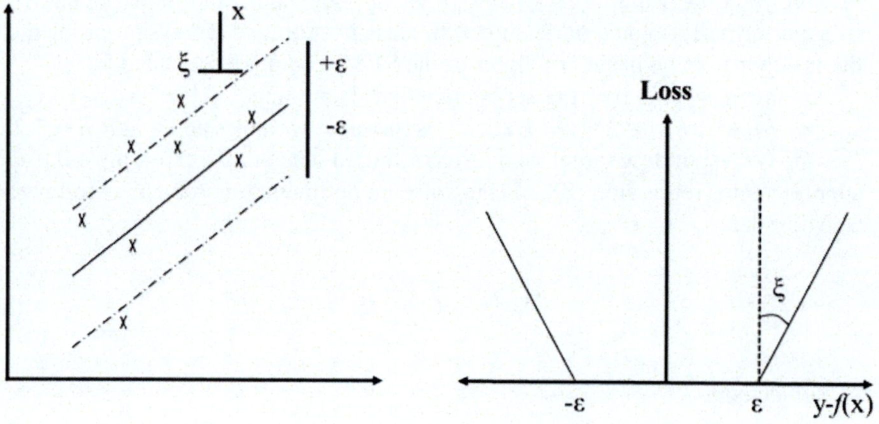

Fig. 3.14 ε-Insensitive loss function and slack variable ξ in SV-regression (after Samui and Dixon 2012)

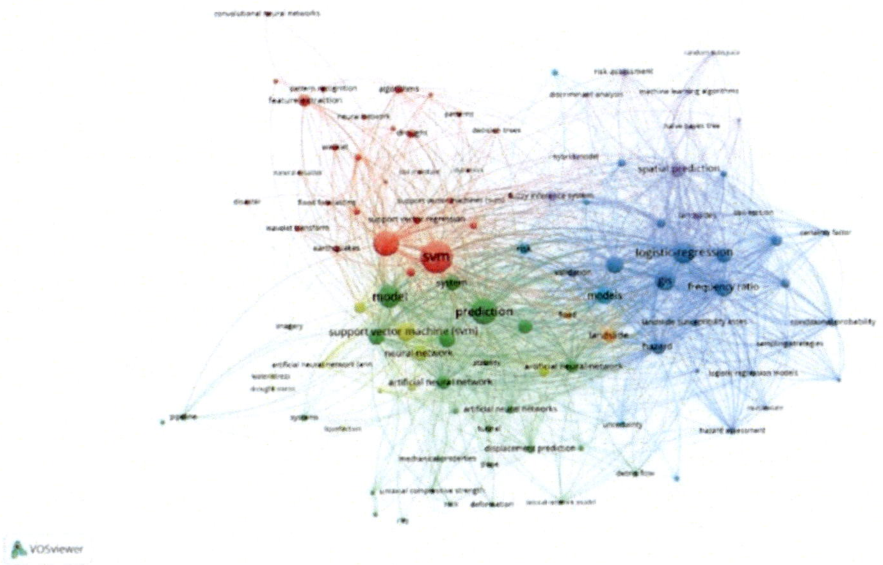

Fig. 3.15 Visualization of SVM algorithm

the hyperparameters to obtain an optimal model to give the accurate predictions with minimum errors. To avoid overfitting, a technique called k-fold cross-validation is widely used. For more details about SVM, it can refer to Samui and Dixon (2012); Xue et al. (2014); Zhang et al. (2016). Figure 3.15 provides the Visualization of SVM algorithm.

3.1.8 KNN

KNN is the abbreviation for K-nearest neighbor, one kind of basic supervised learning algorithm. KNN can be adopted to solve both classification and regression problems (Wu et al. 2008). By comparison, the classification problem is more widely used. Therefore, algorithms for classification are mainly introduced here. The data set is first divided into training and testing sets. The training set is utilized to preliminarily establish the KNN model and then the validation set is used to obtain the appropriate hyperparameters. Then, the testing set is adopted to verify the established model. The key principle for KNN classification is to classify the new data into the nearest classes according to the Euclidean Distance, Manhattan Distance or Minkowski Distance and the equations to calculate these three kinds of distances are listed in Eqs. (3.9, 3.10 and 3.11), respectively. Figure 3.16 provides the Visualization of KNN algorithm.

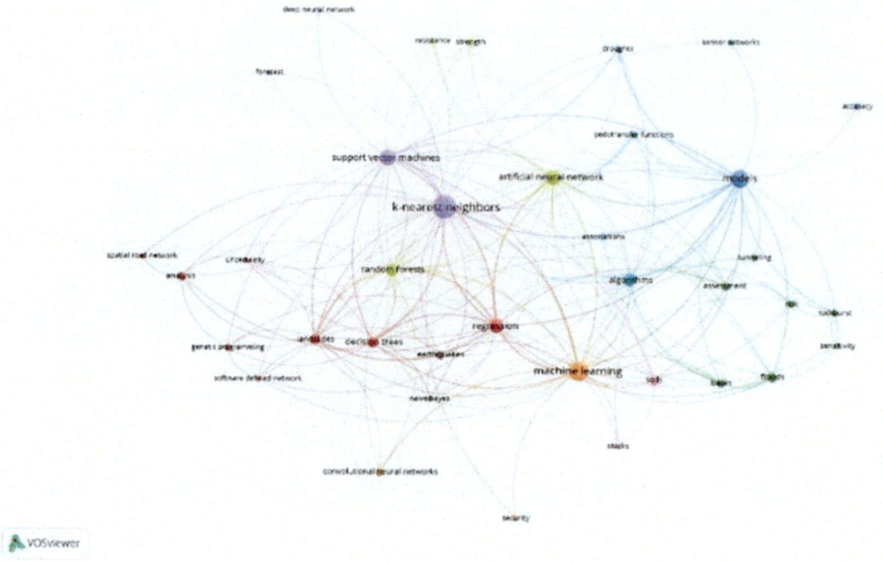

Fig. 3.16 Visualization of KNN algorithm

$$\mathrm{D}_{Euclidean} = \sqrt{\sum_{i=1}^{k} (x_i - y_i)^2} \tag{3.9}$$

$$D_{Manhattan} = \sum_{i=1}^{k} |x_i - y_i| \tag{3.10}$$

$$D_{Minkowski} = \left\{ \sum_{i=1}^{k} (|x_i - y_i)^q \right\}^{1/q} \tag{3.11}$$

3.2 Unsupervised Learning

Unsupervised learning uses ML algorithms to analyze and cluster unlabeled datasets. These algorithms discover hidden patterns or data groupings without the need for human intervention. Its ability to discover similarities and differences in information make it the ideal solution for exploratory data analysis, cross-selling strategies, customer segmentation, image and pattern recognition. It's also used to reduce the number of features in a model through the process of dimensionality reduction; principal component analysis (PCA) and singular value decomposition (SVD) are two common approaches for this. Other algorithms used in unsupervised learning include neural networks, k-means clustering, probabilistic clustering methods, and more.

3.2.1 K-means

K-means is a typical unsupervised learning algorithm, which divides the sample into several disjoint clusters (Jain 2010). The clustering results have high intra-cluster similarity and low inter-cluster similarity. The idea of k-means algorithm is relatively simple. Assuming that the data is to be divided into K categories, the steps are as follows:

The k-means algorithm first initializes k different centroids $\{\mu^{(1)}, \cdots, \mu^{(k)}\}$, calculates the distance between each sample and the centroids k, and selects the cluster with the closest centroid $\mu^{(i)}$ as the category of sample i. Then recalculating the centroid of each cluster and adjusting the category of each sample again until reaching the maximum number of iterations or the adjustment range is less than the threshold. Since the similarity between all samples and each centroid is calculated each time, the convergence speed of the k-means algorithm is relatively slow on large-scale data sets.

3.2.2 PCA

In the field of data science, we are frequently faced with the problem of overfitting. Reducing the dimension of the original feature space and simplifying structures hidden in the database is an efficient solution. In order to reduce the dimensionality, there are two commonly used techniques, namely, feature elimination and feature extraction. Feature elimination is used to drop out unimportant variables that contribute less to the target. It is a simple method, and it remains the interpretability as possible. Feature extraction is used to generate new independent variables that account for the correlation in the database, and each newly created variable is a combination of the original variables. Principal components analysis (PCA) is a widely used technique of feature extraction. It combines the input variables in a specific way, and the most important independent principal components can then be used as predictor variables for further analysis. It has several strengths; however, the model interpretability is deteriorated as original variables are not remained at all (Bakshi 1998).

Figure 3.17 provides the VOSviewer plots for some main unsupervised learning methods. For brevity, the specific explanations as for the size of the node and the word, as well as the distance are omitted.

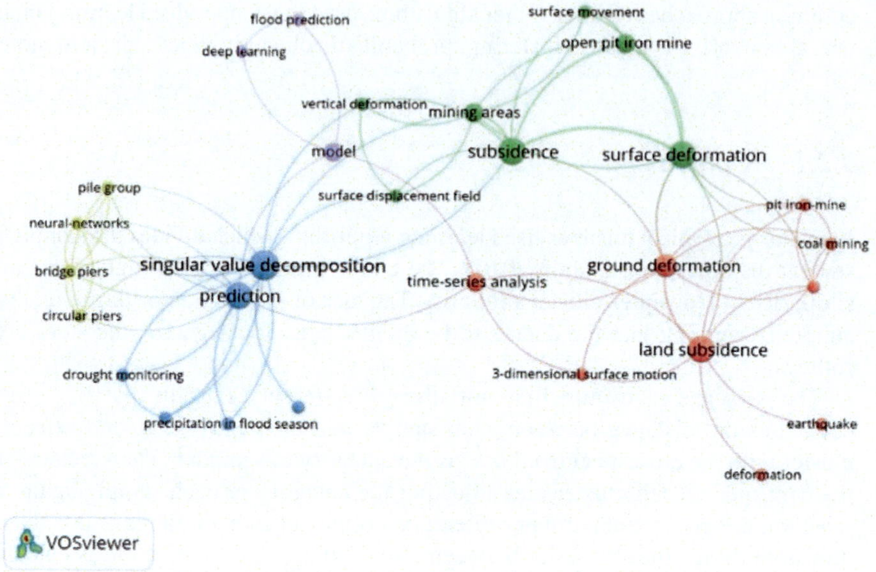

(a)

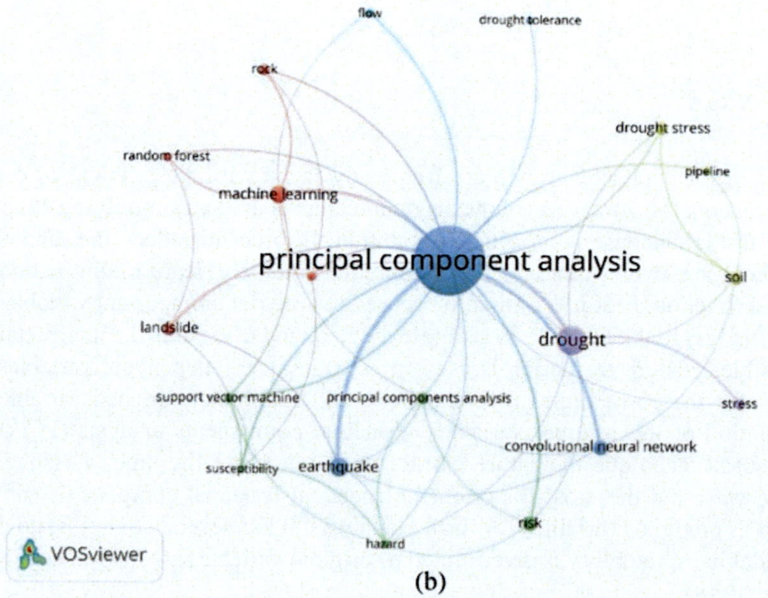

(b)

Fig. 3.17 Network visualization of **a** SVD **b** PCA **c** K means **d** Probabilistic clustering

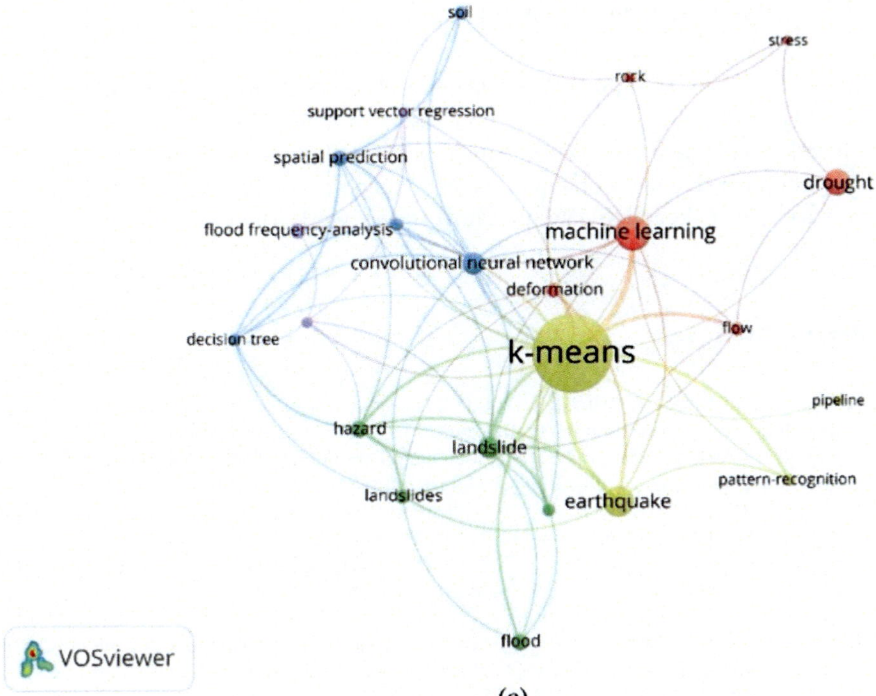

(c)

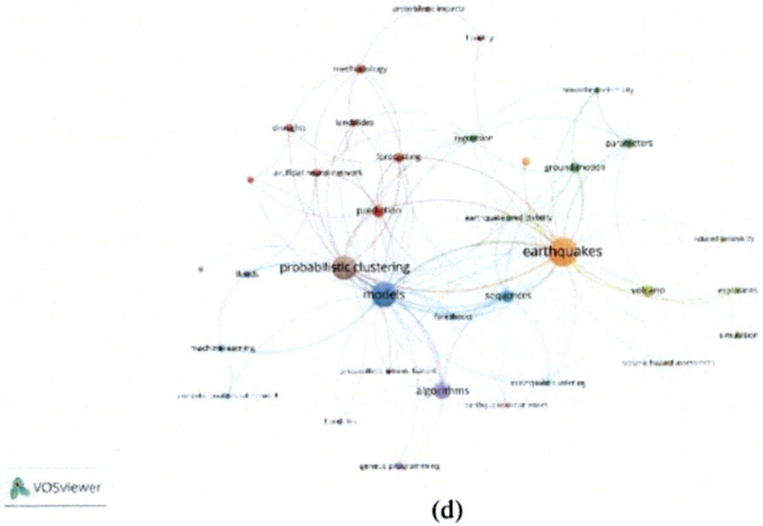

(d)

Fig. 3.17 (continued)

3.3 Semi-Supervised Learning

Semi-supervised learning offers a happy-to-both medium between supervised and unsupervised learning. During training, it uses a smaller labeled data set to guide classification and feature extraction from a larger, unlabeled data set. Semi-supervised learning can solve the problem of having not enough labeled data (or not being able to afford to label enough data) to train a supervised learning algorithm.

References

Agatonovic-Kustrin S, Beresford R (2000) Basic concepts of artificial neural network (ANN) modeling and its application in pharmaceutical research. J Pharm Biomed Anal 22:717–727
Angeline PJ (1994) Genetic programming: On the programming of computers by means of natural selection. Biosystems 33:69–73. https://doi.org/10.1016/0303-2647(94)90062-0
Bakshi BR (1998) Multiscale PCA with application to multivariate statistical process monitoring. AIChE J 44:1596–1610. https://doi.org/10.1002/aic.690440712
Ben Chaabene W, Flah M, Nehdi ML (2020) Machine learning prediction of mechanical properties of concrete: Critical review. Constr Build Mater 260https://doi.org/10.1016/j.conbuildmat.2020.119889
Bhargavi P, Jyothi S (2009) Applying Naive Bayes data mining technique for classification of agricultural land soils. IJCSNS Int J Comput Sci Netw Secur 9:117–122
Bin HG, Zhu QY, Siew CK (2006) Extreme learning machine: Theory and applications. Neurocomputing 70:489–501. https://doi.org/10.1016/j.neucom.2005.12.126
Breiman L, Friedman JH, Olshen RA, Stone CJ (2017) Classification and regression trees. Routledge
Elmousalami HH (2020) Artificial intelligence and parametric construction cost estimate modeling: state-of-the-art review. J Constr Eng Manag 146:03119008. https://doi.org/10.1061/(asce)co.1943-7862.0001678
Jain AK (2010) Data clustering: 50 years beyond K-means. Pattern Recognit Lett 31:651–666. https://doi.org/10.1016/j.patrec.2009.09.011
Koza JR (1992) The genetic programming paradigm: Genetically breeding populations of computer programs to solve problems. Dyn Genet Chaotic Program 203–321. https://doi.org/10.1109/TAI.1990.130444
Murthy SK (1998) Automatic construction of decision trees from data: A multi-disciplinary survey. Data Min Knowl Discov 2https://doi.org/10.1023/A:1009744630224
Niuniu X (2010) Review of decision trees. 2010 3rd Int Conf Comput Sci Inf Technol 105–109
Oltean M, Groşan C, Dioşan L, Mihăilă C (2009) Genetic programming with linear representation: A survey. Int J Artif Intell Tools 18
Pham BT, Tien Bui D, Pourghasemi HR et al (2017) Landslide susceptibility assesssment in the Uttarakhand area (India) using GIS: a comparison study of prediction capability of naïve bayes, multilayer perceptron neural networks, and functional trees methods. Theor Appl Climatol 128:255–273. https://doi.org/10.1007/s00704-015-1702-9
Quinlan JR (1993) {C4}.5—Programs for Machine Learning
Rashidi HH, Tran NK, Betts EV, et al (2019) Artificial intelligence and machine learning in pathology: The present landscape of supervised methods. Acad Pathol 6
Samui P, Dixon B (2012) Application of support vector machine and relevance vector machine to determine evaporative losses in reservoirs. Hydrol Process 26:1361–1369. https://doi.org/10.1002/hyp.8278
Vapnik VN (2000) The nature of statistical learning theory. Nat Stat Learn Theory. https://doi.org/10.1007/978-1-4757-3264-1

Wei W, Ramalho O, Malingre L et al (2019) Machine learning and statistical models for predicting indoor air quality. Indoor Air 29:704–726
Wieczorek W, Czech ZJ (2000) Grammars in genetic programming. Control Cybern 29:1018–1030
Wu X, Kumar V, Ross QJ et al (2008) Top 10 algorithms in data mining. Knowl Inf Syst 14:1–37. https://doi.org/10.1007/s10115-007-0114-2
Xue XH, Yang XG, Chen X (2014) Application of a support vector machine for prediction of slope stability. Sci China Technol Sci 57:2379–2386. https://doi.org/10.1007/s11431-014-5699-6
Zhang Y, Dai M, Ju Z (2016) Preliminary discussion regarding SVM kernel function selection in the twofold rock slope prediction model. J Comput Civ Eng 30:04015031. https://doi.org/10.1061/(asce)cp.1943-5487.0000499

Chapter 4
Deep Learning and Applications

DL attempts to mimic the human brain, albeit far from matching its ability, enabling systems to cluster data and make predictions with incredible accuracy. It consist of multiple layers of interconnected nodes, each building upon the previous layer to refine and optimize the prediction or categorization. This progression of computations through the network is called forward propagation. The input and output layers of a deep neural network are called visible layers. The input layer is where the deep learning model ingests the data for processing, while the output layer is where the final prediction or classification is made.

Another process called backpropagation uses algorithms, like gradient descent, to calculate errors in predictions and then adjusts the weights and biases of the function by moving backwards through the layers in an effort to train the model. Together, forward propagation and backpropagation allow a neural network to make predictions and correct for any errors accordingly. Over time, the algorithm becomes gradually more accurate. The above describes the simplest type of deep neural network in the simplest terms. However, DLs are incredibly complex, and there are different types of neural networks to address specific problems or datasets, for example, AE, DBN, CNN, RNN and LSTM, etc.

4.1 AE

Autoencoder is mainly used in dimensionality reduction and compression (Hinton and Zemel 1994). The strategy is to try to make the output equal to the input or attempts to reconstruct the data. The automatic encoder includes an encoder and a decoder. The encoder receives an input, then encodes it into a vector in a low-dimensional latent space, and then the decoder is responsible for decoding the vector to obtain the original input.

AEs can obtain a feature representation of a less dimensional input from the output in the network, which achieves dimensionality reduction and compression.

W. Zhang et al., *Application of Soft Computing, Machine Learning, Deep Learning and Optimizations in Geoengineering and Geoscience*,
https://doi.org/10.1007/978-981-16-6835-7_4

In addition, it can be utilized to retrieve slightly different input data, or even better data, which can be used for training data enhancement, data denoising, etc.

4.2 DBN

Restricted Boltzmann machine (RBM) is a stochastic neural network with generating ability, which can learn probability distribution through input (LeCun et al. 1989; Hinton 1989). Compared with other networks, its biggest feature is that it has only input and hidden layers, not output. In the forward training, a corresponding feature representation generated after an input passed in, and then in the backpropagation, the original input is reconstructed from this feature representation. This process is similar to autoencoder, but it realized in a single network.

Multiple restricted Boltzmann machines can be stacked to form a DBN. It is similar to the fully connected layer, but the training way is different. The training of DBN is to train its network layers in pairs according to the training process of RBMs. Recently, fewer people use DBN and RBM due to the emergence of generative adversarial networks and mutated autoencoders.

4.3 CNN

CNN is a kind of deep learning algorithm that has been widely used for video recognition, image classification, recommendation systems and self-driving problems (LeCun et al. 1998; Krizhevsky et al. 2017). Compared with traditional methods, CNN has unparallel prediction accuracy and working efficiency, as the pre-processing stage requires less efforts. The core feature of CNN lies in the building block which is composed of many layers that are interconnected with each other using a set of shared weights and biases. Figure 4.1 shows a typical architecture of a CNN model, containing convolutional, pooling and fully connected layers.

4.4 RNN

Recurrent neural network is a dynamic neural network and can capture the time-series data (Graves et al. 2013). As shown in Fig. 4.2, the basic structure of RNN is demonstrated. The advantage of this algorithm is to be able to capture the sequential information. To be specific, a former unit may have influence on the next unit. Based on this structure, the information will pass from one layer to the next sequence by sequence.

The input of RNN can be multiple and orderly, and it can transfer the information of the upper hidden layer to the next hidden layer of the neuron. That's to say, the

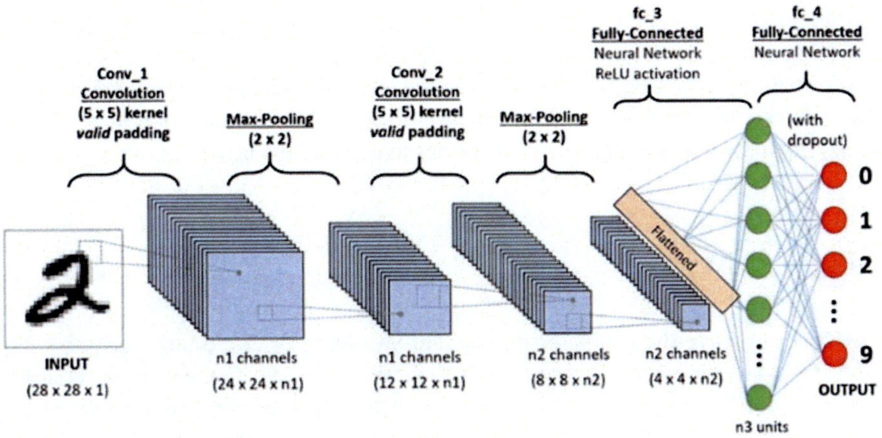

Fig. 4.1 Model structure of CNN structure

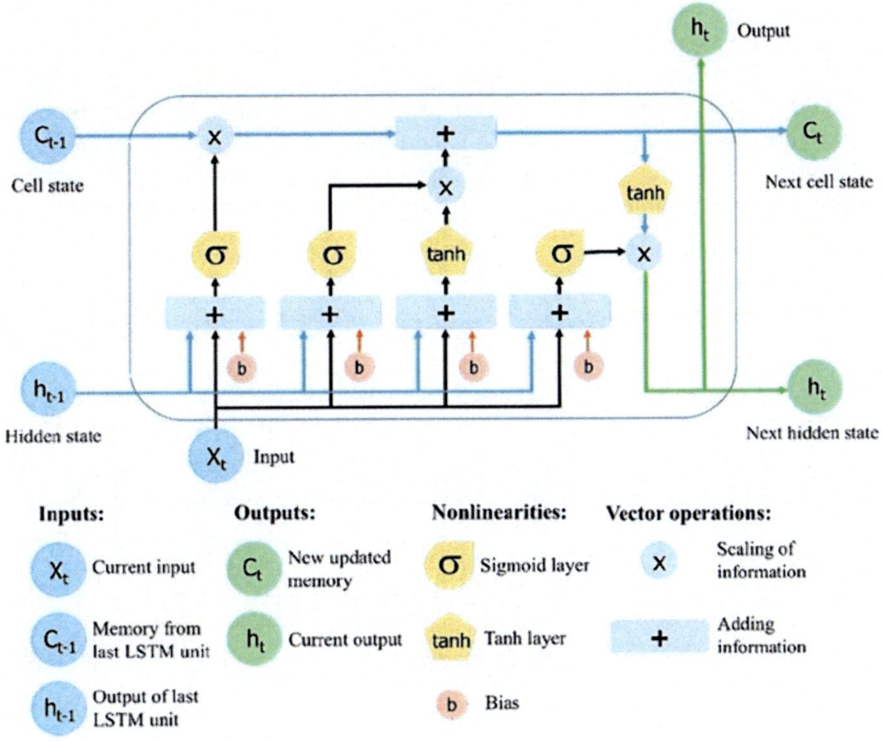

Fig. 4.2 Model structure of RNN structure (after Le et al. 2019)

RNN algorithm has a certain memory ability. Furthermore, RNN only remembers the important information and then definitely forgets the remaining unimportant information. Although the principles of the RNN algorithm are straightforward and simple, yet it may be difficult to obtain an ideal RNN model when there is a long time-lag between the target and its former associated events.

4.5 LSTM

In order to solve the problem of 'gradient vanishing or exploding' faced by RNN, LSTM was proposed (Hochreiter and Schmidhuber 1997). Compared with RNN, LSTM makes better use of historical information for sequential problems, and it is much more applicable for complex sequence learning problems. The main improvement for LSTM lies in the 'memory cell', as illustrated in Fig. 4.3.

The memory cell in LSTM contains three specially designed gates, namely input gate, forget gate, and output gate. The input gate controls how much new information should be passed to the cell state. The forget gate determines what information will be remained or removed from the cell state. The output gate calculates outcome results, and this process is generally carried out by a sigmoid function. In a word, based on a combination of the current input and the previous output, the LSTM model predicts the target value at next time step. This process continues throughout the whole life cycle before the required error or the maximal iteration number is reached.

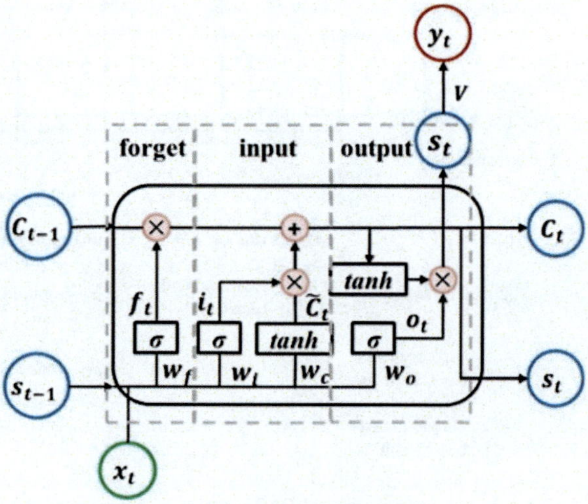

Fig. 4.3 Model structure of LSTM structure (Zhao et al. 2019)

References

Graves A, Mohamed AR, Hinton G (2013) Speech recognition with deep recurrent neural networks. In: ICASSP, IEEE International conference on acoustics, speech and signal processing—proceedings, pp 6645–6649

Hinton GE (1989) Deterministic Boltzmann learning performs steepest descent in weight-space. Neural Comput 1:143–150. https://doi.org/10.1162/neco.1989.1.1.143

Hinton GE, Zemel RS (1994) Autoencoders, minimum description length and Helmholtz free energy. Adv Neural Inf Process Syst 6:3–10

Hochreiter S, Schmidhuber J (1997) Long short-term memory. Neural Comput 9:1735–1780. https://doi.org/10.1162/neco.1997.9.8.1735

Krizhevsky A, Sutskever I, Hinton GE (2017) ImageNet classification with deep convolutional neural networks. Commun ACM 60:84–90. https://doi.org/10.1145/3065386

Le D-T, Lauw HW, Fang Y (2019) Correlation-sensitive next-basket recommendation. Proceedings of the Twenty-Eighth International Joint Conference on Artificial Intelligence (IJCAI-19). 2808–2814.

LeCun Y, Bottou L, Bengio Y, Haffner P (1998) Gradient-based learning applied to document recognition. Proc IEEE 86:2278–2323. https://doi.org/10.1109/5.726791

LeCun Y, Galland CC, E HG (1989) GEMINI: Gradient estimation through matrix inversion after noise injection

Zhao H, Chen Z, Jiang H, et al (2019) Evaluation of three deep learning models for early crop classification using sentinel-1A imagery time series—a case study in Zhanjiang, China. Remote Sens 11:2673. https://doi.org/10.3390/RS11222673

Chapter 5
Optimization Algorithms and Applications

When you're trying to make tough decisions about questions that involve an inordinate number of factors, optimization helps you to capture key components to build a mathematical model of the engineering situation, giving you the confidence to make better decisions more quickly.

Optimization problems can be divided into function optimization problems and combinatorial optimization problems. The object of function optimization problem is a continuous variable in a certain interval, which solves the problem of continuity. The object solution space of combinatorial optimization is discrete state, i.e., it solves discrete problems. Typical combinatorial optimization includes Traveling salesman problem (TSP), Scheduling problem, Knapsack problem, Bin packing problem, Graph coloring problem, Clustering problem, etc. Optimization problems can also be divided into precise algorithm and approximation algorithm (heuristic algorithm). The first is generally rather complex, only suitable for solving small-scale problems, and not practical in geo engineering. The latter is based on intuitionistic or empirical construction which gives a feasible solution of each instance of the combinatorial optimization problem to be solved under acceptable time and space cost. The deviation between the feasible solution and the optimal solution cannot be predicted. Metaheuristic algorithm is an improvement of heuristic algorithm. It is the combination of random algorithm and local search algorithm. The metaheuristic algorithm an also be further divided as: individual-based, such as Tabu search algorithm and simulated annealing, and population-based, such as evolutionary algorithm and novel swarm intelligence optimization mentioned above. Nowadays, the use of swarm intelligence optimization in geoengineering and geo-science is dominant.

W. Zhang et al., *Application of Soft Computing, Machine Learning, Deep Learning and Optimizations in Geoengineering and Geoscience*,
https://doi.org/10.1007/978-981-16-6835-7_5

5.1 Precise Algorithm

As mentioned in Fig. 1.2 about the basic components of OA algorithms, the precise algorithm consists of the Stochastic Gradient Descent, Conjugate Gradient (CG) as well as the steepest Descent. Figure 5.1 provides the VOSviewer plots for SGD and CG use in geoengineering and geosciences.

5.2 Evolutionary Algorithm

Evolutionary computation is an approach to engineering and optimization in which solutions, instead of being constructed from first principles, are instead evolved through processes modeled after the elements of Darwinian evolution. Evolutionary computation is one of the principal methods in what is called 'nature-inspired computing.' The evolutionary algorithm is the main object of interest in evolutionary computation. There is a problem to be solved, and the solution is conceived to lie somewhere in a space of possible candidate solutions—the search space. Based on the literature survey, the common evolutionary algorithms include:

5.2.1 GA

GA is one of the most representative strategies for simulating evolutionary computing, which proposed by Holland (1992). It is a random global search and optimization method that imitates the evolution of natural organisms. This method acquires and accumulates knowledge about the search space during the search process, and adaptively controls the search process to obtain the optimal solution. It mainly adopts crossover and mutation operations to achieve global search, which effectively reduces the disadvantages of traditional algorithms that are easy to fall into local optimal solutions. The standard genetic algorithm consists of encoding process, fitness function and genetic operators. Figure 5.2 provides the network Visualization of GA algorithm.

5.2.2 BO

Snoek et al.(2012) proposed BO algorithm for hyperparameter tuning in 2012. Since its excellent capabilities and the advantages of time saving, it has been widely used in deep learning, system optimization, environmental monitoring, etc., which providing a new direction for solving complex optimization problems.

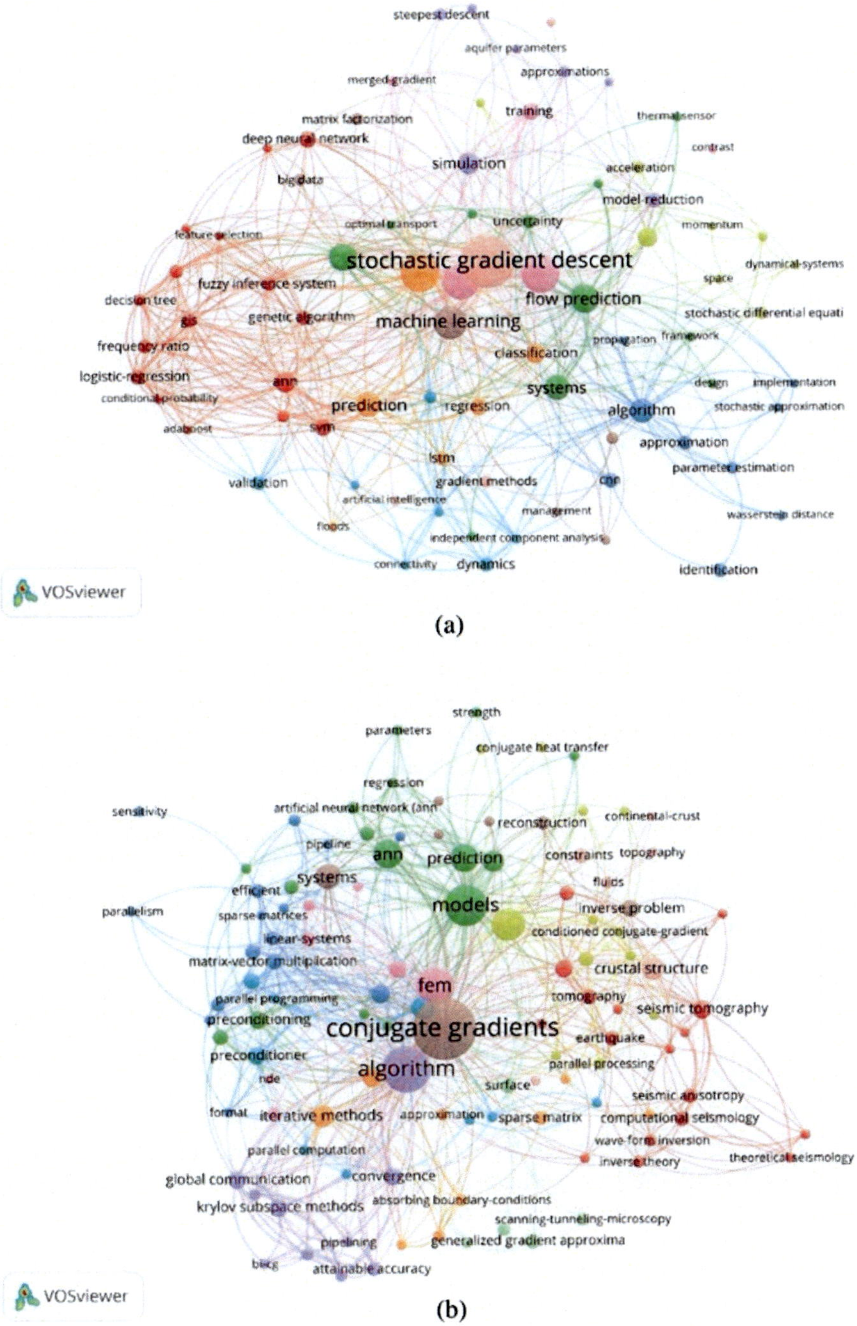

Fig. 5.1 VOSviewer plots for **a** SGD and **b** CG use in geoengineering and geosciences

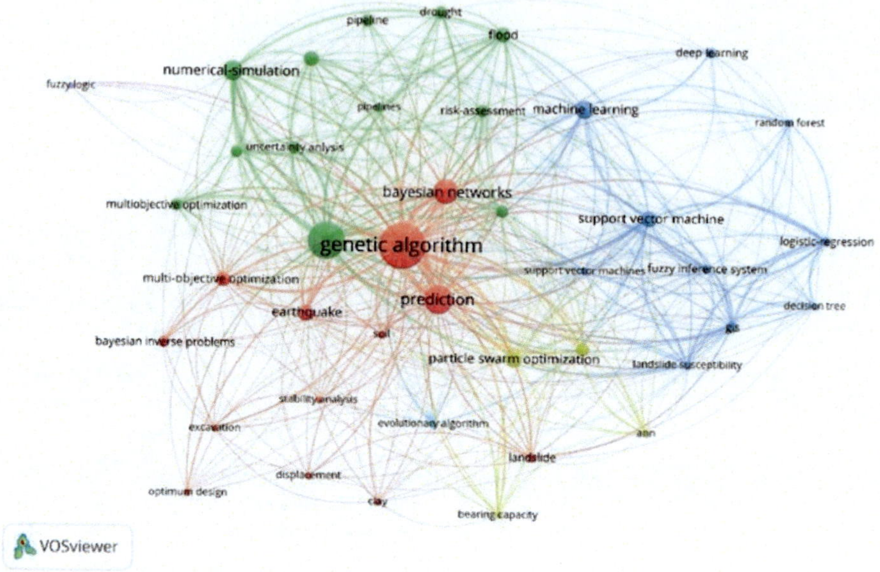

Fig. 5.2 Network visualization of GA algorithm

BO can be regarded as a process of finding the optimal parameters to make the unknown objective function to reach the maximum or minimum value. This process involves the two core parts of the Bayesian optimization algorithm, namely the probabilistic surrogate model and the acquisition function. BO algorithm uses a probabilistic surrogate model to fit the true objective function, and actively selects the evaluation point with the most potential for the next evaluation based on the acquisition function, thereby the optimal solution of the objective function can be obtained using a small number of evaluation times, and avoid unnecessary sampling. Figure 5.3 provides the network Visualization of BO algorithm.

5.2.3 *PSO*

PSO is a global optimization algorithm proposed by Eberhart and Kennedy (1995). The main idea is to treat particles as each solution of the optimization problem, and each particle swims in groups based on flight experience of its own and other particles. In order to measure the superiority of each particle solution, it defines the fitness value function to search for the optimal solution from the whole space. Figure 5.4 provides the Network Visualization of PSO algorithm.

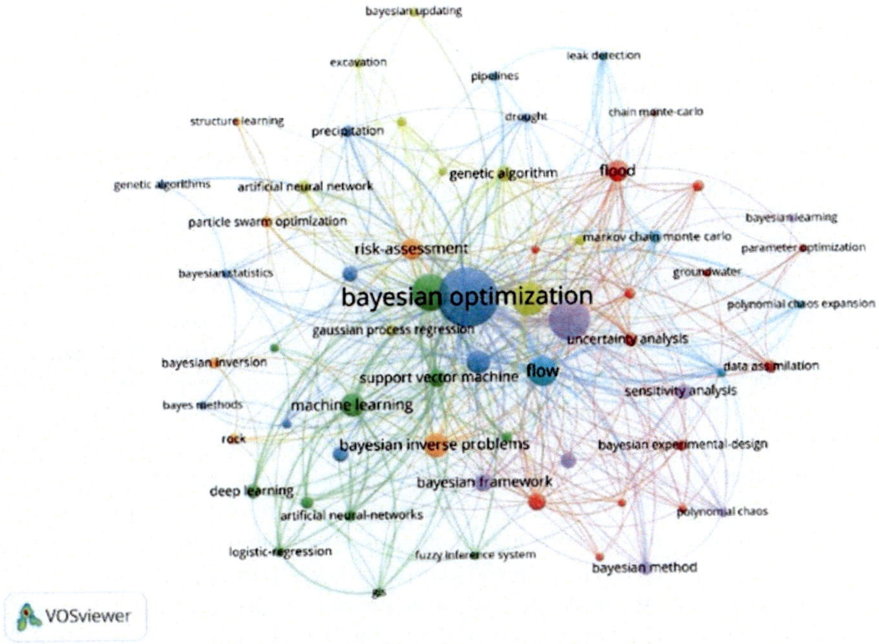

Fig. 5.3 Network visualization of BO algorithm

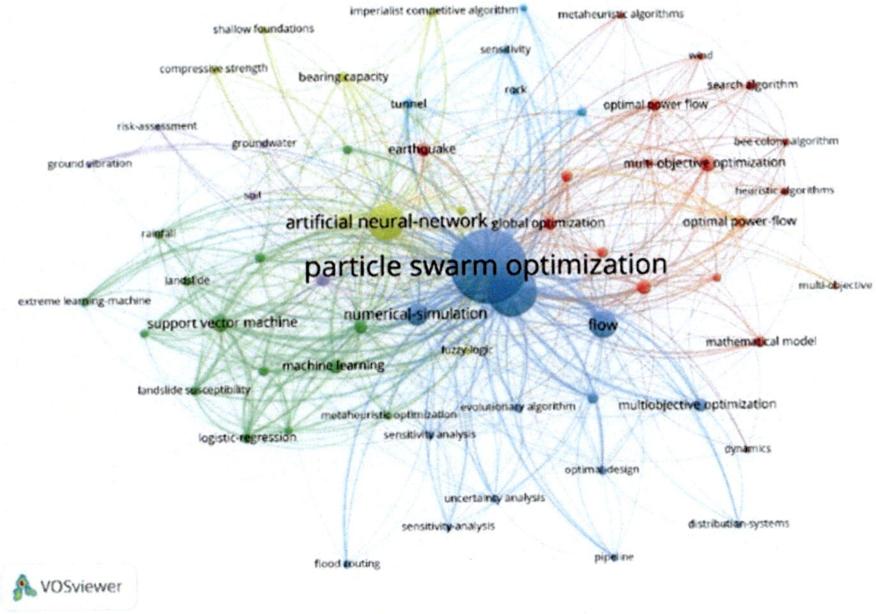

Fig. 5.4 Network visualization of PSO algorithm

5.2.4 DE

DE is a stochastic, simple, and efficient evolutionary algorithm to find global optima for numerous unconstrained test functions by Storn and Price (1997). Compared with other evolutionary algorithms, DE reserves population-based global search strategy and uses a simple mutation operation of the differential and one-on-one competition, so it can reduce the genetic complexity of the operation (Huang and Chen 2013). Meanwhile, the special memory capacity of DE makes it available to dynamically track the current search to adjust the search strategy with a strong global convergence and robustness. Therefore, it is suitable for solving some complex optimization problems (Huang and Chen 2013). The main procedure of DE contains three main operators: namely, mutation, crossover, and selection. In an iterative process, assume that the population of each generation G contains N individuals. The initial population is often randomly generated following a uniform distribution over the variable domain. The mutation operation is used to generate an individual, by adding the weighted difference between two population individuals to a third one. The crossover operation is introduced to diverse the new population. According to the crossover strategy, the old and new individual exchange part of the code to form a new individual. The selection operation is greedy strategy, which compares the cost function values of a candidate individual and the target individual to determine the best one. Figure 5.5 provides the Network Visualization of DE algorithm.

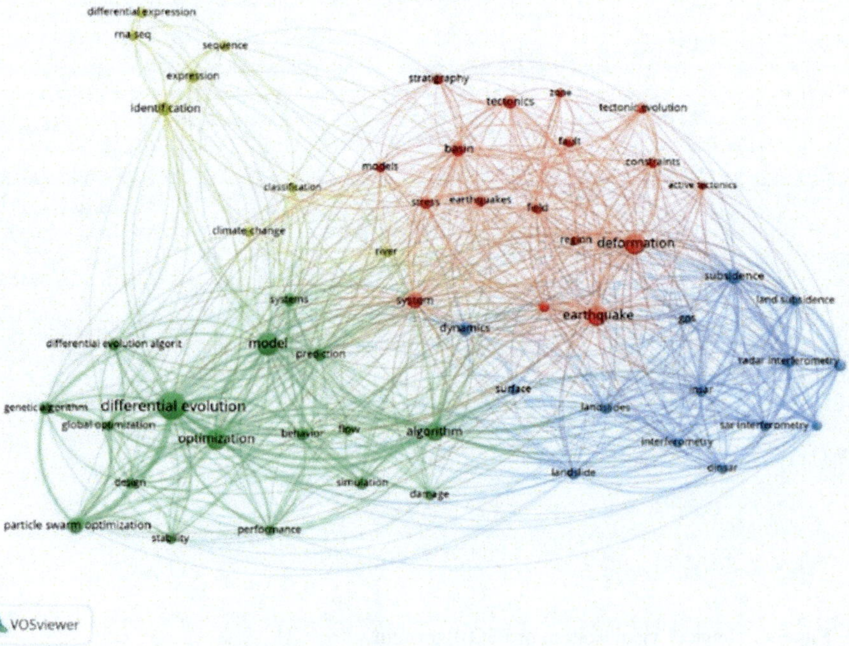

Fig. 5.5 Network visualization of DE algorithm

5.2.5 ABC

ABC is a swarm-based meta-heuristic algorithm to solve optimization problems. It is inspired by the intelligent foraging behavior of swarm honeybees (Karaboga 2005). In the ABC algorithm, there are three types of bees in the colony: employed bees (forager bees), onlooker bees (observe bees) and scouts. Each type of bee handles a specific task. Employed bees exploit and search for food sources, and for each food source there is only one employed bee. Employed bees share information with the onlooker bees in a hive so that onlooker bee can choose a food source to forager. Scouts are associated with searching for new food sources by exploring the environment surrounding the hive. The waggle dance is used to show the nectar content of the food source explored by a dancing bee, and the duration of the dance is proportional to quality of food source. Onlooker bees will watch numerous dances before choosing a good food source according to the potential quality of that food source. When a bee, whether it is scout or onlooker, finds a food source, it becomes employed. When a food source is discarded, the employed bees associated it will again become scouts or onlookers randomly. It is found that scouts are performing the job of exploration, while employed and onlooker bees are performing the job of exploitation (Garg 2014). There are three important controlling parameters in ABC, which are colony size, limit, and maximum cycle (Yusup et al. 2014). A more detailed calculation process to achieve this algorithm can be referred to (Karaboga and Akay 2009). Figure 5.6 provides the Network Visualization of ABC algorithm.

5.2.6 ACO

ACO is a recently developed, population-based technique applied for solving optimization problems by Dorigo and Gambardella (1997). This algorithm is inspired by the foraging behavior of real ant colonies. When searching for food, the ants explore the environment surrounding their nest in a random manner. If an ant finds a food source, it will evaluate it and carry some food back to the nest. During the return trip, the ant marks the path by laying some amount of pheromone on the edge of the path. The pheromone contains information of quantity and quality of the food source, and other ants are then attracted by the pheromone leading them to the food. The pheromone trails represent an indirect communication way (also known as stigmergy) among ants, making it available to find shortest path between the nest and food source (Nait Amar et al. 2018). Figure 5.7 provides the Network Visualization of ACO algorithm.

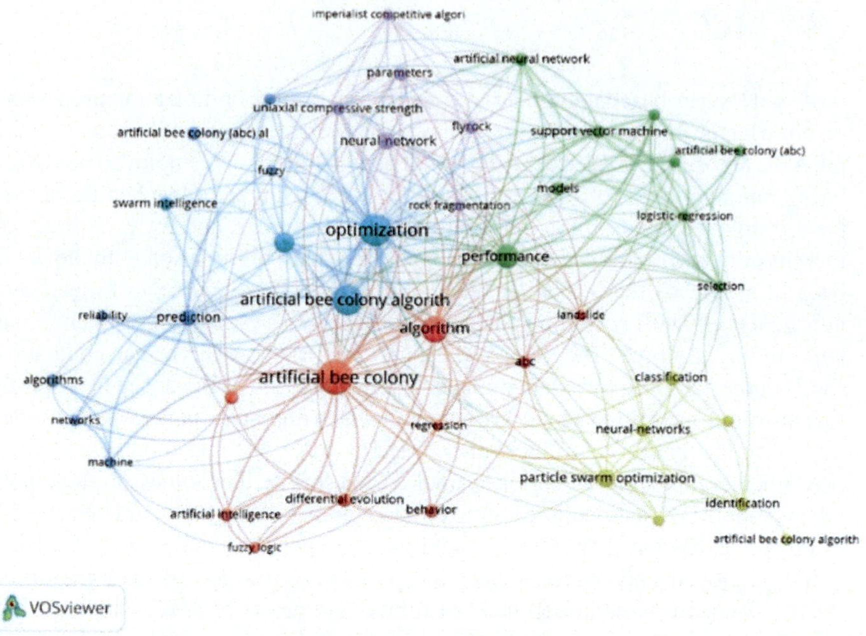

Fig. 5.6 Network visualization of ABC algorithm

5.2.7 CS

CS is a nature-inspired population-based stochastic metaheuristic algorithm developed by Yang and Deb (2009). The main idea of this algorithm is inspired by the parasitic behavior of some cuckoo species in combination with the Lévy flight, which is a motion way used by birds to search for food. In the nature, some cuckoos lay their eggs in the nests of other species of birds with amazing abilities selecting suitable nests where host birds recently laid eggs and removing existing eggs to increase the hatching probability of their own eggs. The cuckoo search algorithm can be described based on three-point idealization conditions (Kanagaraj et al. 2014; Zhang et al. 2019). A more detailed construction process to achieve this algorithm can be referred to Yang and Deb (2009). Figure 5.8 provides the Network Visualization of CS algorithm.

5.2.8 FA

FA is a heuristic novel bionic optimization algorithm (Yang 2009), which was inspired by the flickering behavior of fireflies, using points in search space to simulate firefly individuals in nature. FA can deal with multi-modal functions more efficiently than

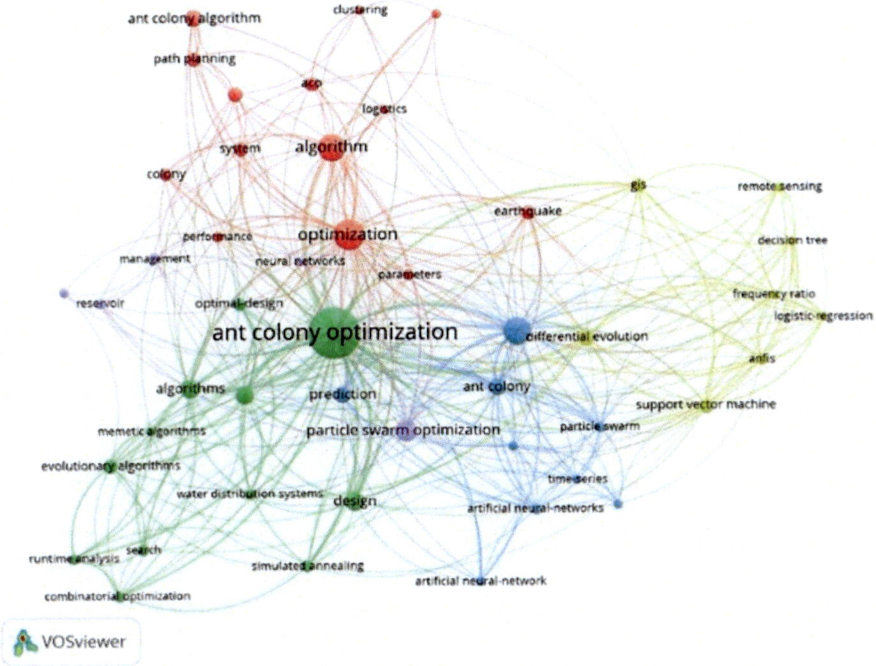

Fig. 5.7 Network visualization of ACO algorithm

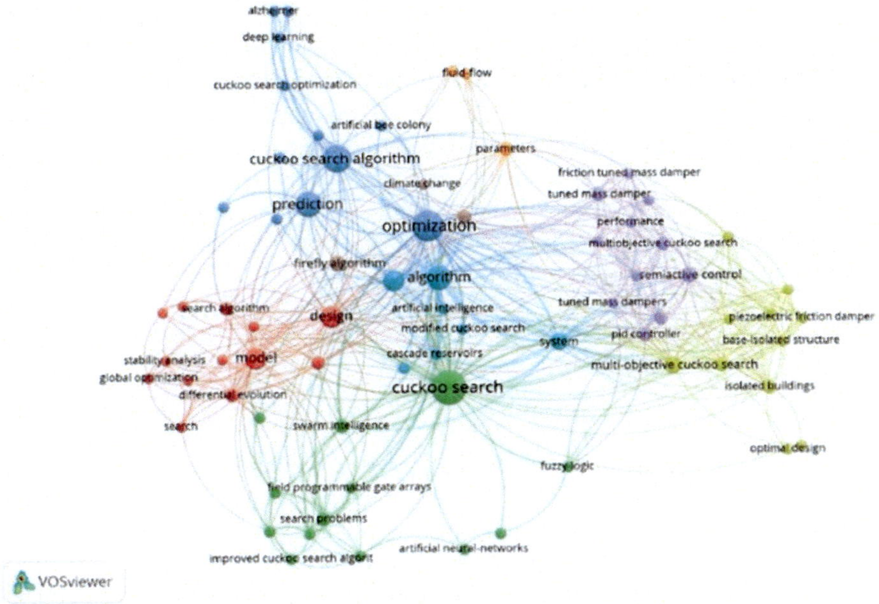

Fig. 5.8 Network visualization of CS algorithm

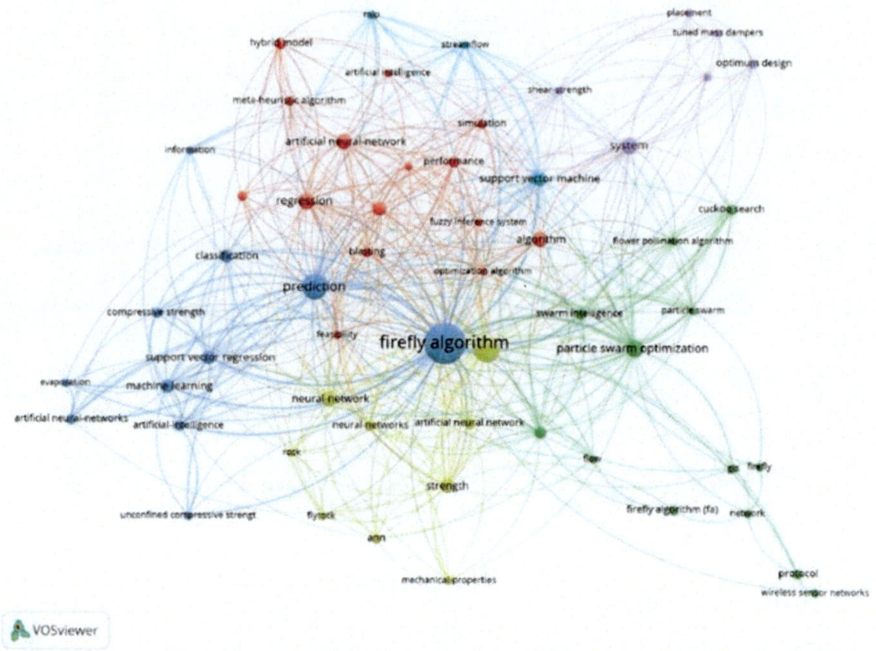

Fig. 5.9 VOSviewer plots for FA algorithm use in geoengineering and geosciences

other swarm algorithms (Łukasik and Zak 2009; Yang 2009; Gandomi et al. 2011; Yang et al. 2012). The search process is simulated as the attraction and movement process of firefly individuals. FA mainly includes two basic elements: brightness and attractive brightness (Huang et al. 2019). Brightness mainly reflects the quality of the location of the firefly which can determine the direction in which the firefly moves. Attraction determines the distance the fireflies can travel. The optimization of the goal is done by constantly updating the position of the fireflies in brightness and attractiveness. Figure 5.9 provides the VOSviewer plots for FA Algorithm use in geoengineering and geosciences.

5.2.9 GWO

Inspired by the predation behavior of gray wolves, (Mirjalili et al. 2014) proposed a new type of swarm intelligence optimization algorithm: GWO algorithm in 2014. GWO achieves the goal of optimization by simulating the predation behavior of gray wolves, based on the cooperation mechanism of wolves. Figure 5.10 provides the details of GWO, including the demonstration of hunting behavior, i.e., tracking, chasing, and approaching the prey; pursuing, encircling, and harassing the prey until it

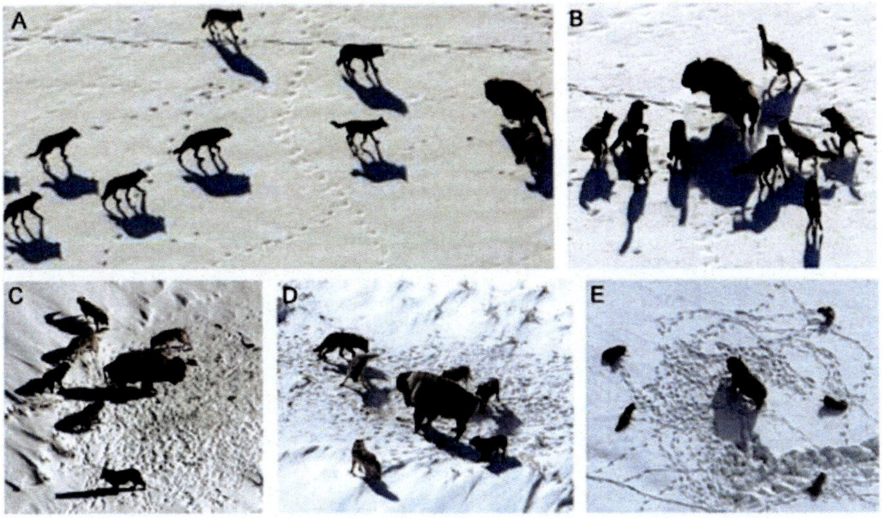

(a)

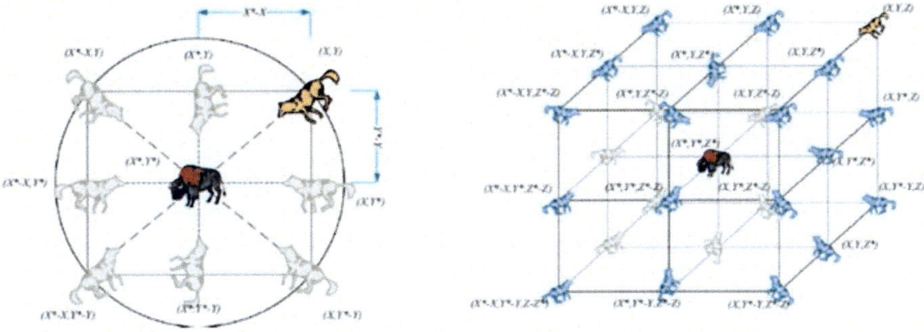

(b)

Fig. 5.10 GWO: **a** demonstration of hunting behavior, **b** 2D and 3D position vectors, **c** position updating and **d** basic flowchart

Fig. 5.10 (continued)

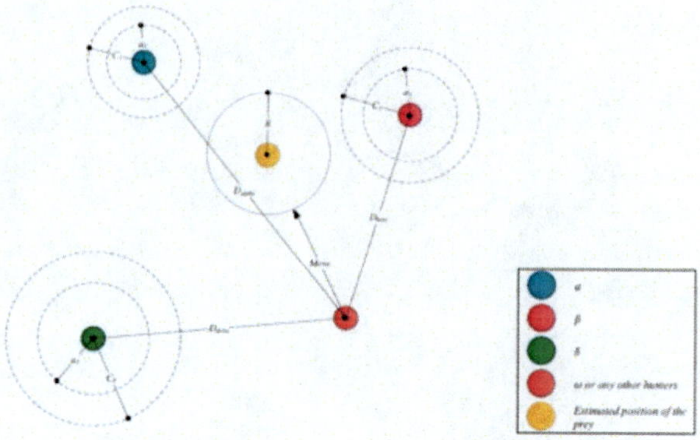

(c)

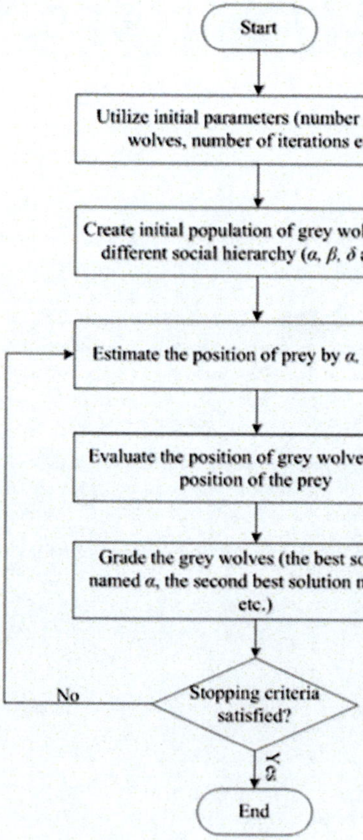

(d)

Fig. 5.11 The Whale optimization algorithm (after Mirjalili and Lewis (2016))

stops moving; attack towards the prey, 2D and 3D position Vectors, position updating and basic flowchart.

5.2.10 WOA

Mirjalili and Lewis (2016) introduced and developed Whale Optimization Algorithm, which is one of the metaheuristic algorithms and proved its capability in balancing between the exploration and exploitation compared to the state-of-the-art optimization algorithms. The humpback whales use the bubble-net hunting technique (Aljarah et al. 2018). WOA algorithm is composed of three phases including: (1) encircling prey phase; (2) exploitation phase; (3) exploration phase (search for a prey) (Fig. 5.11).

5.2.11 SFLA

SFLA Eusuff and Lansey (2003) is a meta-heuristic to solve discrete optimization problems (Elbeltagi et al. 2005). It originated in the feeding behavior of frogs. The Memetic Algorithm (MA) is a swarm intelligence algorithm used by Moscato (1989) to solve optimization problems through heuristic search. "Memetie" means that it is

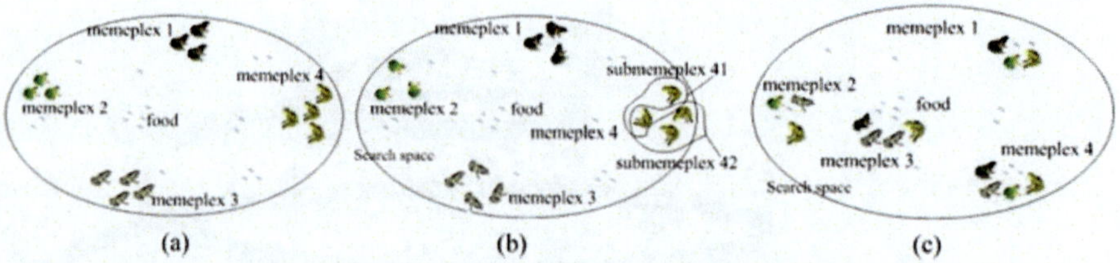

Fig. 5.12 Shuffled Frog leaping algorithm (after Zhu and Zhang (2014))

stored in the human or animal brain to guide its behavior and spread. The information is similar to the chromosomes in genes. The SFLA, in essence, combines the benefits of the genetic-based MAs and the social behavior-based PSO algorithms. Figure 5.12 depicts the working mechanism of SFLA.

5.2.12 CSO

CSO is a swarm Intelligence algorithm, which was originally invented by Chu et al. (2006). It is inspired by the natural behavior of cats, and it has a novel technique in modeling exploration and exploitation phases. It has been successfully applied in various optimization fields of science and engineering. However, the literature lacks a recent and detailed review of this algorithm, especially in geoengineering and geosciences. For brevity, the summary of CSO application is omitted. Figure 5.13 shows the sketch map and flowchart of CSO. For simplicity, the basic concept consists of (1) "seeking mode" for when cats are resting and (2) "tracing mode" for when they are chasing their prey.

Figure 5.14 provides the pie chart statistics for applications of GWO, WOA, and SFLA in geoengineering and geosciences.

Table 5.1 lists the corresponding applications of these three methods mentioned in the relevant papers.

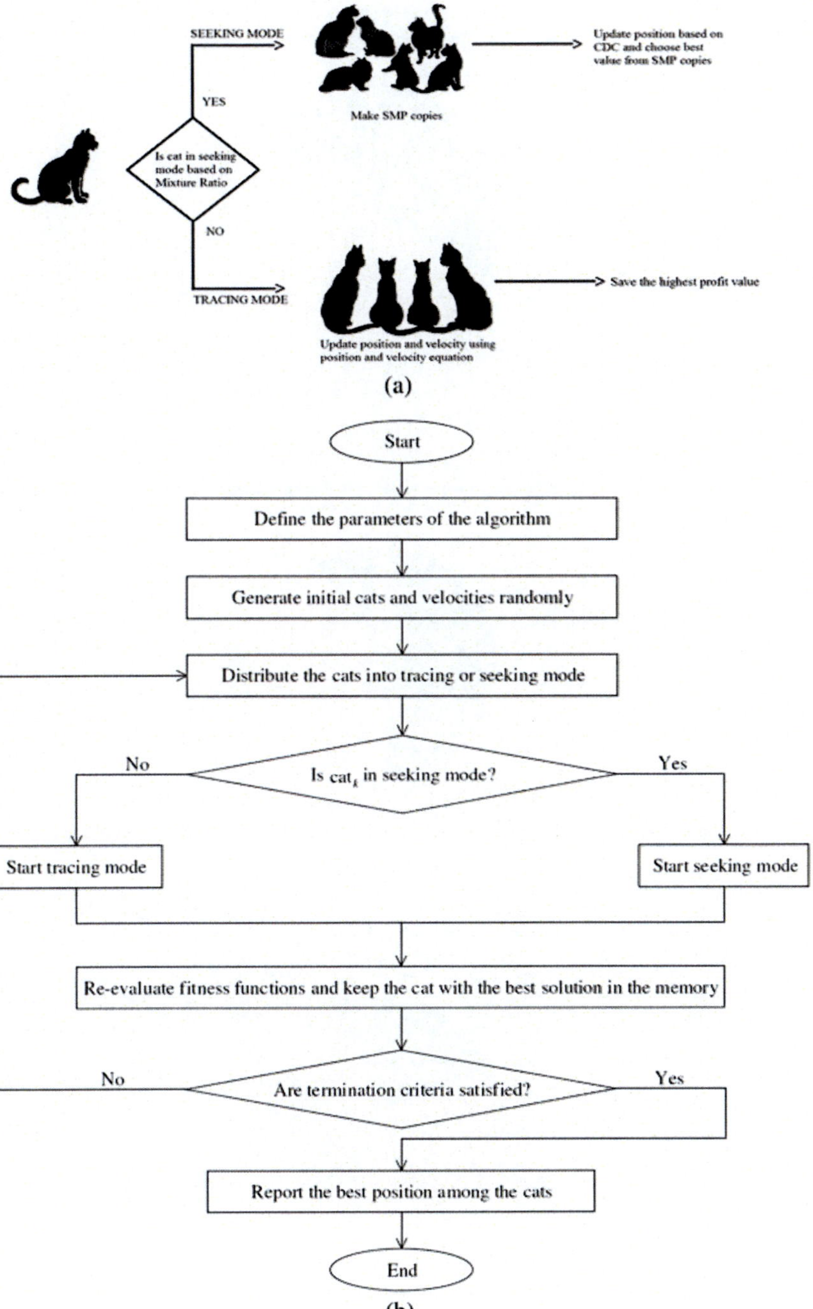

Fig. 5.13 Sketch map (**a**) and flowchart (**b**) of CSO (partially after Selvakumar et al. 2017)

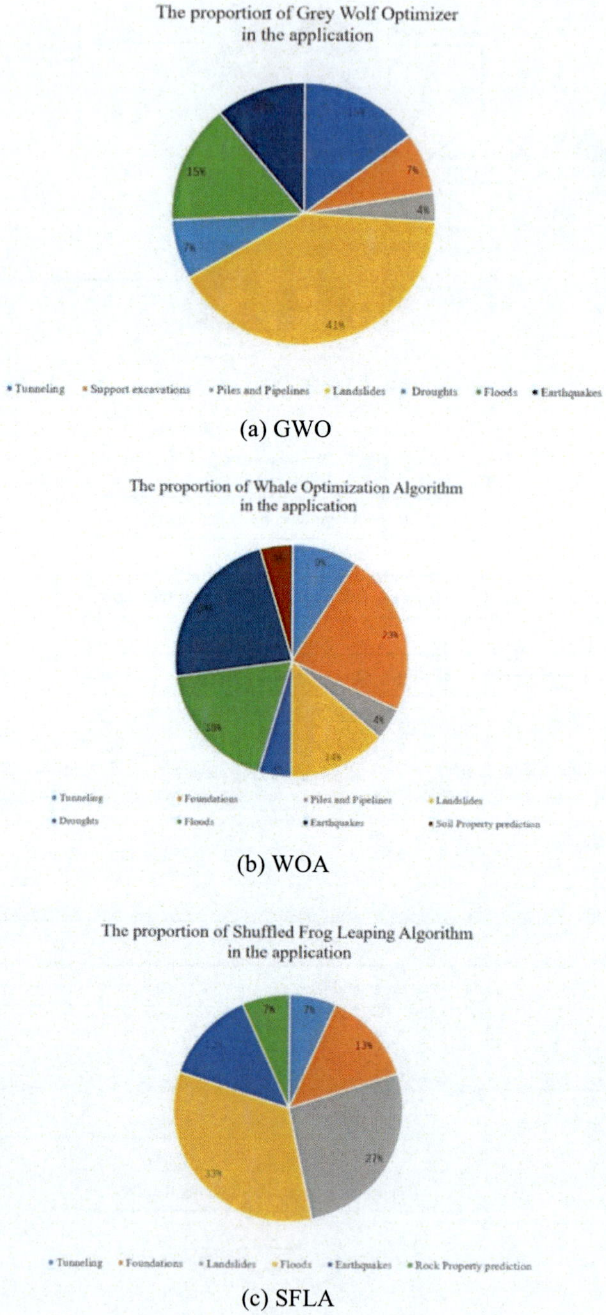

Fig. 5.14 Pie chart statistics for applications of **a** GWO, **b** WOA and **c** SFLA

Table 5.1 Most relevant papers mentioning application of GWO, WOA and SFLA in geoengineering and geosciences

Methods	References	Objective	Use
GWO	Chen and Dong (2020)	Tunneling	Optimize an accurate and stable water intake forecasting model
GWO	Li et al. (2021a)	Tunneling	Predict the penetration rate
GWO	Yagiz and Karahan (2015)	Tunneling	Estimate the performance of TBM
GWO	Yang et al. (2020)	Tunneling	Optimize the most important parameters of TBM performance
GWO	Gandomi and Kashani (2018)	Foundations	Deal with shallow optimization in two ways
GWO	Temur and Oser (2019)	Foundations	Research the robustness of pile group optimization problem under vertical load
GWO	Dehghanbanadaki et al. (2021)	Piles and Pipelines	Confirmed the effectiveness and high accuracy of the MLP-GWO model
GWO	Temur and Oser (2019)	Piles and Pipelines	Calculate the bearing capacity of a single pile and estimate the optimal number of piles in the group
GWO	Abdollahizad et al. (2021)	Landslides	Landslide sensitivity mapping framework
GWO	Chen et al. (2019a)	Landslides	Landslide modeling framework
GWO	Guo et al. (2020)	Landslides	Develop a new prediction model GWO-BP
GWO	Jaafari et al. (2019)	Landslides	Estimate landslide sensitivity
GWO	Jiang et al. (2021)	Landslides	Predict the displacement of a landslide and use the prediction interval to quantify its uncertainty
GWO	Kadirhodjaev et al. (2020)	Landslides	Generate a landslide sensitivity map in a specific area of Uzbekistan
GWO	Li et al. (2021b)	Landslides	Generate debris flow sensitivity map
GWO	Panahi et al. (2020)	Landslides	Develop a landslide sensitivity map for Icheon Town, South Korea

(continued)

Table 5.1 (continued)

Methods	References	Objective	Use
GWO	Paryani et al. (2020)	Landslides	Landslide susceptibility maps were generated by ANFIS-GWO and ANFIS-PSO hybrid models
GWO	Pham et al. (2020)	Landslides	Landslide sensitivity mapping in Van Chan District, Am Bai Province, Vietnam
GWO	Malik et al. (2021)	Droughts	Meteorological drought (MD) prediction
GWO	Tikhamarine et al. (2019)	Droughts	Improve the accuracy of the SVR for predicting monthly traffic
GWO	Mustaffa et al. (2018)	Floods	Predict the water level one week in advance
GWO	Rahmati et al. (2020)	Floods	Develop a new hybrid model based on historical urban flood events
GWO	Zou et al. (2019)	Floods	Search for the optimal fuzzy clustering center and the sensitivity coefficient of FCI-CW globally
GWO	Kamgar et al. (2018)	Earthquakes	Control the seismic vibration of the structure
GWO	Vijay and Nanda (2019)	Earthquakes	Identify and extract clustering events
GWO	Zhou et al. (2021a)	Earthquakes	Predict the potential of soil liquefaction
GWO	Temur and Oser (2019)	Support excavations	Optimization of pile groups under vertical loads
WOA	Li et al. (2021c)	Tunneling	Verify the applicability of the proposed hybrid GEP and WOA method
WOA	Zhou et al. (2021b)	Tunneling	Optimizing the hyper-parameters of the SVM technique in forecasting TBM AR
WOA	Lin et al. (2020)	Foundations	Analyze the time-related work behavior of RCCD
WOA	Moayedi et al. (2019)	Foundations	Analyze the bearing capacity of two-layer soils

(continued)

Table 5.1 (continued)

Methods	References	Objective	Use
WOA	Zhong et al. (2020)	Foundations	Predict and control cement intake
WOA	Ye et al. (2020b)	Piles and Pipelines	Estimate the physical behavior of the enlarged pile
WOA	Chen et al. (2019a)	Landslides	An integrated landslide modeling framework proposed
WOA	Li et al. (2020)	Landslides	Locate the critical slip surface of a soil slope
WOA	Wei et al. (2021)	Landslides	FOS evaluation and analysis of slopes in two different stages
WOA	Yan et al. (2021)	Droughts	Estimate daily ET0 at four stations in the arid region and four stations in the humid region of China
WOA	Anaraki et al. (2021)	Floods	Flood frequency analysis under climate change conditions
WOA	Jafari-Asl et al. (2021)	Floods	Propose a new framework for the probabilistic design of the labyrinth spillway structure
WOA	Liu et al. (2020)	Floods	Solve the ambiguity problem in resilience evaluation
WOA	Samantaray and Sahoo (2021)	Floods	Predict SSC at Sundargarh and Salebhata stations in the Mahanadi River, India
WOA	Ashkezari and Yazdi (2021a)	Earthquakes	Solve the complex problem of the location allocation of rescue centers
WOA	Ashkezari and Yazdi (2021b)	Earthquakes	Solve the location problem of the earthquake relief center
WOA	Brandao and Miguel (2020)	Earthquakes	Reduce the top layer response and inter-layer drift to a value lower than the limit of inter-layer drift allowed by the reference standard specification
WOA	Lara-Valencia et al. (2021)	Earthquakes	Improve the seismic safety of the structure

(continued)

Table 5.1 (continued)

Methods	References	Objective	Use
WOA	Vaheddoost et al. (2020)	Soil Property prediction	Define the FC and the PWP at the basin scale
SFLA	Mora-Melia et al. (2015)	Tunneling	Size of the water pipe network determination
SFLA	Ye et al. (2020a)	Foundations	Stability analysis using novel intelligent models
SFLA	Zhong et al. (2019)	Foundations	A prediction model for maximum ground settlement of foundation pit proposed
SFLA	Abdollahizad et al. (2021)	Landslides	Efficient landslide susceptibility mapping framework
SFLA	Chen et al. (2019b)	Landslides	Produce a novel ensemble data mining technique
SFLA	Sun et al. (2021)	Landslides	Predict valley deformation
SFLA	Wang et al. (2015)	Landslides	Optimization algorithm of landslide treatment schemes selection proposed
SFLA	Bozorg-Haddad et al. (2015)	Floods	Estimation of Muskingum parameters
SFLA	Ferdowsi et al. (2020)	Floods	Find the best shape of a dam's spillway under climate change impacts
SFLA	Li et al. (2010)	Floods	A novel multi-objective shuffled frog leaping algorithm proposed
SFLA	Orouji et al. (2013)	Floods	Determine optimal/near-optimal parameters in the Muskingum model
SFLA	Shirzadi et al. (2020)	Floods	Predict flash flood susceptibility at Haraz, Iran
SFLA	Lin and Chen (2016)	Earthquakes	Establish optimal relationship between seismic responses and command voltages of MR dampers
SFLA	Lin et al. (2018)	Earthquakes	Search for the optimal weighting functions
SFLA	Fattahi and Bazdar (2017)	Rock Property prediction	Build a prediction model for the indirect estimation of DRI

References

Abdollahizad S, Balafar MA, Feizizadeh B et al (2021) Using hybrid artificial intelligence approach based on a neuro-fuzzy system and evolutionary algorithms for modeling landslide susceptibility in East Azerbaijan Province, Iran. Earth Sci Inf. https://doi.org/10.1007/s12145-021-00644-z

Aljarah I, Faris H, Mirjalili S (2018) Optimizing connection weights in neural networks using the whale optimization algorithm. Soft Comput 22. https://doi.org/10.1007/s00500-016-2442-1

Anaraki MV, Farzin S, Mousavi SF, Karami H (2021) Uncertainty analysis of climate change impacts on flood frequency by using hybrid machine learning methods. Water Resour Manage 35:199–223. https://doi.org/10.1007/s11269-020-02719-w

Ashkezari AZ, Yazdi HM (2021a) Location allocation of earthquake relief centers in Yazd city based on whale optimization algorithm. Int J Eng 34:1184–1194. https://doi.org/10.5829/ije.2021.34.05b.12

Ashkezari AZ, Yazdi HM (2021b) Presenting a new method for earthquake relief center location allocation based on whale optimization algorithm. Int J Eng 34:769–781. https://doi.org/10.5829/ije.2021.34.04a.03

Bozorg-Haddad O, Hamedi F, Orouji H et al (2015) A re-parameterized and improved nonlinear Muskingum model for flood routing. Water Resour Manage 29:3419–3440. https://doi.org/10.1007/s11269-015-1008-9

Brandao FD, Miguel LFF (2020) Vibration control in buildings under seismic excitation using optimized tuned mass dampers. Frat Ed Integrita Strutt 66–87. https://doi.org/10.3221/igf-esis.54.05

Chen S, Dong S (2020) A sequential structure for water inflow forecasting in coal mines integrating feature selection and multi-objective optimization. IEEE Access 8:183619–183632. https://doi.org/10.1109/access.2020.3028959

Chen W, Hong HY, Panahi M et al (2019a) Spatial prediction of landslide susceptibility using GIS-based data mining techniques of ANFIS with whale optimization algorithm (WOA) and grey wolf optimizer (GWO). Appl Sci 9. https://doi.org/10.3390/app9183755

Chen W, Panahi M, Tsangaratos P et al (2019b) Applying population-based evolutionary algorithms and a neuro-fuzzy system for modeling landslide susceptibility. CATENA 172:212–231. https://doi.org/10.1016/j.catena.2018.08.025

Chu SC, Tsai PW, Pan JS (2006) Cat swarm optimization. In: Proceedings of the Pacific Rim International Conference on Artificial Intelligence, pp 854–858. Springer, Guilin, China

Dehghanbanadaki A, Khari M, Amiri ST, Armaghani DJ (2021) Estimation of ultimate bearing capacity of driven piles in c-φ soil using MLP-GWO and ANFIS-GWO models: a comparative study. Soft Comput 25:4103–4119. https://doi.org/10.1007/S00500-020-05435-0

Dorigo M, Gambardella LM (1997) Ant colony system: a cooperative learning approach to the traveling salesman problem. IEEE Trans Evol Comput 1:53–66. https://doi.org/10.1109/4235.585892

Eberhart R, Kennedy J (1995) New optimizer using particle swarm theory. In: Proceedings of the International symposium on micro machine and human science. IEEE, pp 39–43

Elbeltagi E, Hegazy T, Grierson D (2005) Comparison among five evolutionary-based optimization algorithms. Adv Eng Inf 19:43–53. https://doi.org/10.1016/j.aei.2005.01.004

Eusuff MM, Lansey KE (2003) Optimization of water distribution network design using the shuffled frog leaping algorithm. J Water Resour Plan Manage 129:210–225. https://doi.org/10.1061/(asce)0733-9496(2003)129:3(210)

Fattahi H, Bazdar H (2017) Applying improved artificial neural network models to evaluate drilling rate index. Tunn Undergr Sp Technol 70:114–124. https://doi.org/10.1016/j.tust.2017.07.017

Ferdowsi A, Mousavi SF, Farzin S, Karami H (2020) Optimization of dam's spillway design under climate change conditions. J Hydroinformatics 22:916–936. https://doi.org/10.2166/hydro.2020.019

Gandomi AH, Kashani AR (2018) Construction cost minimization of shallow foundation using recent swarm intelligence techniques. IEEE Trans Ind Inf 14:1099–1106. https://doi.org/10.1109/TII.2017.2776132

Gandomi AH, Yang XS, Alavi AH (2011) Mixed variable structural optimization using Firefly algorithm. Comput Struct 89:2325–2336. https://doi.org/10.1016/j.compstruc.2011.08.002

Garg H (2014) Solving structural engineering design optimization problems using an artificial bee colony algorithm. J Ind Manag Optim 10:777–794. https://doi.org/10.3934/jimo.2014.10.777

Guo Z, Chen L, Gui L et al (2020) Landslide displacement prediction based on variational mode decomposition and WA-GWO-BP model. Landslides 17:567–583. https://doi.org/10.1007/S10346-019-01314-4

Holland JH (1992) Adaptation in natural and artificial systems: an introductory analysis with applications to biology, control, and artificial intelligence. Adapt Nat Artif Syst. https://doi.org/10.7551/MITPRESS/1090.001.0001

Huang Y, Li J, Fu J (2019) Review on application of artificial intelligence in civil engineering. Comput Model Eng Sci 121:845–875

Huang Z, Chen Y (2013) An improved differential evolution algorithm based on adaptive parameter. J Control Sci Eng 2013. https://doi.org/10.1155/2013/462706

Jaafari A, Panahi M, Pham BT et al (2019) Meta optimization of an adaptive neuro-fuzzy inference system with grey wolf optimizer and biogeography-based optimization algorithms for spatial prediction of landslide susceptibility. CATENA 175:430–445. https://doi.org/10.1016/j.catena.2018.12.033

Jafari-Asl J, Ben Seghier ME, Ohadi S, van Gelder P (2021) Efficient method using Whale optimization algorithm for reliability-based design optimization of labyrinth spillway. Appl Soft Comput 101. https://doi.org/10.1016/j.asoc.2020.107036

Jiang Y, Xu Q, Lu Z et al (2021) Modelling and predicting landslide displacements and uncertainties by multiple machine-learning algorithms: application to Baishuihe landslide in Three Gorges Reservoir, China. Geomatics Nat Hazards Risk 12:741–762. https://doi.org/10.1080/19475705.2021.1891145

Kadirhodjaev A, Rezaie F, Lee MJ, Lee S (2020) Landslide susceptibility assessment using an optimized group method of data handling model. ISPRS Int J Geoinf 9. https://doi.org/10.3390/ijgi9100566

Kamgar R, Samea P, Khatibinia M (2018) Optimizing parameters of tuned mass damper subjected to critical earthquake. Struct Des Tall Spec Build 27. https://doi.org/10.1002/tal.1460

Kanagaraj G, Ponnambalam SG, Jawahar N, Nilakantan JM (2014) An effective hybrid cuckoo search and genetic algorithm for constrained engineering design optimization. Eng Optim 46:1331–1351. https://doi.org/10.1080/0305215X.2013.836640

Karaboga D, Akay B (2009) A comparative study of Artificial Bee Colony algorithm. Appl Math Comput 214:108–132. https://doi.org/10.1016/j.amc.2009.03.090

Karaboga D (2005) An idea based on Honey Bee Swarm for numerical optimization. Tech Rep TR06, Erciyes Univ 10

Lara-Valencia LA, Caicedo D, Valencia-Gonzalez Y (2021) A novel Whale optimization algorithm for the design of tuned mass dampers under earthquake excitations. Appl Sci 11. https://doi.org/10.3390/app11136172

Li YH, Zhou JZ, Zhang YC et al (2010) Novel multiobjective shuffled Frog leaping algorithm with application to reservoir flood control operation. J Water Resour Plan Manage 136:217–226. https://doi.org/10.1061/(asce)wr.1943-5452.0000027

Li H, Yao H, Chen L et al (2021a) Assessment of circular-bored twin tunnel (CBTT) performance using soft computing methods. Eng Comput. https://doi.org/10.1007/S00366-021-01288-9

Li Y, Chen W, Rezaie F et al (2021b) Debris flows modeling using geo-environmental factors: developing hybridized deep-learning algorithms. Geocarto Int. https://doi.org/10.1080/10106049.2021.1912194

Li Z, Yazdani Bejarbaneh B, Asteris PG et al (2021c) A hybrid GEP and WOA approach to estimate the optimal penetration rate of TBM in granitic rock mass. Soft Comput. https://doi.org/10.1007/S00500-021-06005-8

Li SH, Wu LZ, Luo XH (2020) A novel method for locating the critical slip surface of a soil slope. Eng Appl Artif Intell 94. https://doi.org/10.1016/j.engappai.2020.103733

Lin XF, Chen SM (2016) A modified shuffled frog-leaping algorithm-based fuzzy controller for magnetorheological damper-building system. Int J Comput Appl Technol 53:279–289

Lin X, Chen S, Huang G (2018) A shuffled frog-leaping algorithm based mixed-sensitivity H∞ control of a seismically excited structural building using MR dampers. J Vib Control 24:2832–2852. https://doi.org/10.1177/1077546317695462

Lin CN, Li TC, Chen SY et al (2020) Structural identification in long-term deformation characteristic of dam foundation using meta-heuristic optimization techniques. Adv Eng Softw 148. https://doi.org/10.1016/j.advengsoft.2020.102870

Liu D, Fan ZR, Fu Q et al (2020) Random forest regression evaluation model of regional flood disaster resilience based on the whale optimization algorithm. J Clean Prod 250. https://doi.org/10.1016/j.jclepro.2019.119468

Łukasik S, Zak S (2009) Firefly algorithm for continuous constrained optimization tasks. In: Lecture notes in computer science (including subseries Lecture notes in artificial intelligence and Lecture notes in bioinformatics), pp 97–106

Malik A, Tikhamarine Y, Souag-Gamane D et al (2021) Support vector regression integrated with novel meta-heuristic algorithms for meteorological drought prediction. Meteorol Atmos Phys 133:891–909. https://doi.org/10.1007/S00703-021-00787-0

Mirjalili S, Mohammad S, Lewis A (2014) Advances in engineering software Grey Wolf optimizer. Adv Eng Softw 69

Mirjalili S, Lewis A (2016) The Whale optimization algorithm. Adv Eng Softw 95:51–67. https://doi.org/10.1016/j.advengsoft.2016.01.008

Moayedi H, Bui DT, Ngo PTT (2019) Neural computing improvement using four metaheuristic optimizers in bearing capacity analysis of footings settled on two-layer soils. Appl Sci 9. https://doi.org/10.3390/app9235264

Mora-Melia D, Iglesias-Rey PL, Martinez-Solano FJ, Ballesteros-Pérez P (2015) Efficiency of evolutionary algorithms in water network pipe sizing. Water Resour Manage 29:4817–4831. https://doi.org/10.1007/S11269-015-1092-X

Moscato P (1989) On evolution, search, optimization, genetic algorithms and martial arts: towards memetic algorithms

Mustaffa Z, Sulaiman MH, Ernawan F, Noor NM (2018) Short term forecasting based on hybrid least squares support vector machines. Adv Sci Lett 24:7455–7460. https://doi.org/10.1166/ASL.2018.12958

Nait Amar M, Zeraibi N, Redouane K (2018) Optimization of WAG process using dynamic proxy, genetic algorithm and Ant colony optimization. Arab J Sci Eng 43:6399–6412. https://doi.org/10.1007/s13369-018-3173-7

Orouji H, Haddad OB, Fallah-Mehdipour E, Mariño MA (2013) Estimation of Muskingum parameter by meta-heuristic algorithms. Proc Inst Civ Eng Water Manage 166:315–324. https://doi.org/10.1680/WAMA.11.00068

Panahi M, Gayen A, Pourghasemi HR et al (2020) Spatial prediction of landslide susceptibility using hybrid support vector regression (SVR) and the adaptive neuro-fuzzy inference system (ANFIS) with various metaheuristic algorithms. Sci Total Environ 741. https://doi.org/10.1016/j.scitotenv.2020.139937

Paryani S, Neshat A, Javadi S, Pradhan B (2020) Comparative performance of new hybrid ANFIS models in landslide susceptibility mapping. Nat Hazards 103:1961–1988. https://doi.org/10.1007/S11069-020-04067-9

Pham BT, Phong TV, Trung NT et al (2020) Ensemble modeling of landslide susceptibility using random subspace learner and different decision tree classifiers. Geocarto Int. https://doi.org/10.1080/10106049.2020.1737972

Rahmati O, Darabi H, Panahi M et al (2020) Development of novel hybridized models for urban flood susceptibility mapping. Sci Rep 10. https://doi.org/10.1038/s41598-020-69703-7

Samantaray S, Sahoo A (2021) Prediction of suspended sediment concentration using hybrid SVM-WOA approaches. Geocarto Int. https://doi.org/10.1080/10106049.2021.1920638

Selvakumar K, Vijayakumar K, Boopathi CS (2017) CSO based solution for load kickback effect in deregulated power systems. Appl Sci 7:1127. https://doi.org/10.3390/app7111127

Shirzadi A, Asadi S, Shahabi H et al (2020) A novel ensemble learning based on Bayesian belief network coupled with an extreme learning machine for flash flood susceptibility mapping. Eng Appl Artif Intell 96. https://doi.org/10.1016/j.engappai.2020.103971

Snoek J, Larochelle H, Adams RP (2012) Practical Bayesian optimization of machine learning algorithms. In: Advances in Neural information processing systems, pp 2951–2959

Storn R, Price K (1997) Differential evolution—a simple and efficient heuristic for global optimization over continuous spaces. J Glob Optim 11:341–359. https://doi.org/10.1023/A:1008202821328

Sun M, Xu W, Wang H et al (2021) A novel hybrid intelligent prediction model for valley deformation: a case study in Xiluodu Reservoir Region, China. Comput Mater Contin 66:1057–1074. https://doi.org/10.32604/CMC.2020.012537

Temur R, Oser C (2019) Optimization of pile groups under vertical loads using Grey Wolf optimizer. J Polytech Derg 22:19–32. https://doi.org/10.2339/politeknik.417754

Tikhamarine Y, Souag-Gamane D, Kisi O (2019) A new intelligent method for monthly streamflow prediction: hybrid wavelet support vector regression based on grey wolf optimizer (WSVR-GWO). Arab J Geosci 12. https://doi.org/10.1007/s12517-019-4697-1

Vaheddoost B, Guan YQ, Mohammadi B (2020) Application of hybrid ANN-whale optimization model in evaluation of the field capacity and the permanent wilting point of the soils. Environ Sci Pollut Res 27:13131–13141. https://doi.org/10.1007/s11356-020-07868-4

Vijay RK, Nanda SJ (2019) A quantum Grey Wolf optimizer based declustering model for analysis of earthquake catalogs in an ergodic framework. J Comput Sci 36. https://doi.org/10.1016/j.jocs.2019.07.006

Wang F, Su JY, Wang ZT, Destech Publications I (2015) Optimization of landslide treatment schemes based on the two-dimension IOWA operator and SFL algorithm

Wei W, Li XB, Liu JZ et al (2021) Performance Evaluation of hybrid WOA-SVR and HHO-SVR models with various kernels to predict factor of safety for circular failure slope. Appl Sci 11. https://doi.org/10.3390/app11041922

Yagiz S, Karahan H (2015) Application of various optimization techniques and comparison of their performances for predicting TBM penetration rate in rock mass. Int J Rock Mech Min Sci 80:308–315. https://doi.org/10.1016/j.ijrmms.2015.09.019

Yan SC, Wu LF, Fan JL et al (2021) A novel hybrid WOA-XGB model for estimating daily reference evapotranspiration using local and external meteorological data: applications in arid and humid regions of China. Agric Water Manag 244. https://doi.org/10.1016/j.agwat.2020.106594

Yang XS (2009) Firefly algorithms for multimodal optimization. Lecture notes in computer science (including subseries Lecture notes in artificial intelligence and Lecture notes in bioinformatics). Springer, Berlin, Heidelberg, pp 169–178

Yang XS, Deb S (2009) Cuckoo search via Lévy flights. In: Abraham A, Herrera F, Carvalho A, Pai V (eds) 2009 World congress on nature and biologically inspired computing, NABIC 2009—Proceedings. IEEE, pp 210–214

Yang XS, Hosseini SSS, Gandomi AH (2012) Firefly algorithm for solving non-convex economic dispatch problems with valve loading effect. Appl Soft Comput J 12:1180–1186. https://doi.org/10.1016/j.asoc.2011.09.017

Yang H, Wang Z, Song K (2020) A new hybrid grey wolf optimizer-feature weighted-multiple kernel-support vector regression technique to predict TBM performance. Eng Comput. https://doi.org/10.1007/S00366-020-01217-2

Ye X, Moayedi H, Khari M, Kok Foong L (2020a) Metaheuristic-hybridized multilayer perceptron in slope stability analysis. Smart Struct Syst 26:263–275. https://doi.org/10.12989/SSS.2020.26.3.263

Ye XY, Lyu ZJ, Foong LK (2020b) Hybridized dragonfly, whale and ant lion algorithms in enlarged pile's behavior. Smart Struct Syst 25:765–778. https://doi.org/10.12989/sss.2020.25.6.765

Yusup N, Sarkheyli A, Zain AM et al (2014) Estimation of optimal machining control parameters using artificial bee colony. J Intell Manuf 25:1463–1472. https://doi.org/10.1007/s10845-013-0753-y

Zhang Y, Ge T, Tian W, Liou YA (2019) Debris flow susceptibility mapping using machine-learning techniques in Shigatse area, China. Remote Sens 11. https://doi.org/10.3390/rs11232801

Zhong D, Liu Z, Su Z et al (2020) Predicting and controlling the cement intake during curtain grouting of dam foundations. Eur J Environ Civ Eng. https://doi.org/10.1080/19648189.2020.1847689

Zhong GQ, Wang H, Li L et al (2019) Prediction of maximum settlement of foundation pit based on SFLA-GRNN model. Yantu Lixue/Rock Soil Mech 40. https://doi.org/10.16285/j.rsm.2017.1652

Zhou J, Huang S, Wang M, Qiu Y (2021a) Performance evaluation of hybrid GA–SVM and GWO–SVM models to predict earthquake-induced liquefaction potential of soil: a multi-dataset investigation. Eng Comput. https://doi.org/10.1007/S00366-021-01418-3

Zhou J, Qiu YG, Zhu SL et al (2021b) Optimization of support vector machine through the use of metaheuristic algorithms in forecasting TBM advance rate. Eng Appl Artif Intell 97:. https://doi.org/10.1016/j.engappai.2020.104015

Zhu GY, Zhang WB (2014) An improved Shuffled Frog-leaping algorithm to optimize component pick-and-place sequencing optimization problem. Expert Syst Appl 41:6818–6829. https://doi.org/10.1016/j.eswa.2014.04.038

Zou Q, Liao L, Ding Y, Qin H (2019) Flood classification based on a fuzzy clustering iteration model with combined weight and an immune Grey Wolf optimizer algorithm. Water 11. https://doi.org/10.3390/w11010080

Chapter 6
Application of LSTM and Prophet Algorithm in Slope Displacement Prediction

6.1 Introduction

China is one of the countries with the most serious geological disasters in the world (Wang et al. 2004; Keqiang et al. 2008; Li et al. 2010a; Du et al. 2013). In some cases, landslides can cause huge losses in human lives and properties as well as impose significant impacts on the environment (Yin et al. 2010; Zhang et al. 2020a). Specially, due to the complex geological conditions associated with high mountain areas and intensive human engineering activities, landslides occur frequently in the Three Gorges Reservoir (TGR) region, where steep slopes exist and flooding often occurs (Xu and Niu 2018; Xie et al. 2019). The Qianjiangping landslide on 13 July 2003 destroyed 4 factories and 129 houses, caused 24 deaths, and rendered more than 1000 people homeless (Jian et al. 2014). It is one of the most devastating landslides in the TGR, as it occurred shortly after the first impoundment of the reservoir (Wang et al. 2004; Jian et al. 2014).

It has been well acknowledged that the coupling effect of seasonal rainfall and periodic fluctuation of the reservoir water level poses a great threat to slope safety (Li et al. 2010a; Du et al. 2013; Huang et al. 2017; Yang et al. 2019; Chen et al. 2020; Wang et al. 2020a; Zhang et al. 2020a). However, landslides are complex dynamic and nonlinear systems, and their failure mechanisms have not been fully explored by researchers (Xu and Niu 2018). In recent decades, in order to better control disaster risks and issue early warnings, monitoring networks and controlling systems have been widely used in landslide-prone areas (Li et al. 2010a). Based on the large amount of recorded data, it is possible to analyze the complex relationships between the landslide displacement and its triggering factors. The accuracy of the landslide displacement prediction model depends on how these complex relationships are expressed and reflected (Xu and Niu 2018; Yang et al. 2019).

Conventional landslide displacement approaches such as physical experiments, numerical analysis and empirical formulas have certain advantages yet also disadvantages (Huang et al. 2017; Xu and Niu 2018; Yang et al. 2019). For example, laboratory experiments perform well in revealing the landslide evolution and enhance

W. Zhang et al., *Application of Soft Computing, Machine Learning, Deep Learning and Optimizations in Geoengineering and Geoscience*,
https://doi.org/10.1007/978-981-16-6835-7_6

the understanding on the failure mechanisms, e.g., based on the creep theory and soil and rock mechanics. On the other hand, these experiments are costly and require more time and effort when compared with numerical simulations. Recently, with the development of artificial intelligence, numerous data-driven models have been used for landslide displacement prediction (Li et al. 2010a; Du et al. 2013; Zhou et al. 2016, 2018; Huang et al. 2017; Xie et al. 2019; Yang et al. 2019). Compared with the above methods, the data-driven models are gaining increasing popularity because they achieve a balance between computational costs and model accuracy (Lian et al. 2015; Wang et al. 2020b; Zhang et al. 2020b). Therefore, it can be inferred that a combination of conventional approaches and artificial intelligence methods will be much more applicable for landslide problems in the future.

There are two main categories of artificial intelligence methods applied to the prediction of landslide displacement: the ones considering comprehensive effects of triggering factors and others considering the landslide displacement itself (Xu and Niu 2018). Among these methods, artificial neural networks (ANNs), extreme learning machines (ELMs), and support vector machines (SVMs) are commonly used to predict the landslide displacement (Du et al. 2013; Lian et al. 2015; Zhou et al. 2016, 2018; Cao et al. 2016). For instance, Du et al. (2013) used the back-propagation neural network (BPNN) method to predict the landslide displacement, with the time-series models. Cao et al. (2016) proposed an extreme learning machine model to analyze the displacement characteristics, with a time-series expression. Zhou et al. (2016) compared the predicted displacements of an improved SVM model and the BPNN method, decomposing the displacement time series. These examples illustrate how the combination of time-series analysis and machine learning methods to predict landslide displacement is becoming increasingly popular.

Using a time-series model, the landslide displacement is divided into two main parts: the long-term trend primarily controlled by internal factors (i.e., geological conditions and material properties) and the periodic displacement significantly affected by external factors (i.e., rainfall and reservoir water level). The long-term trend is more time-dependent and grows gently, and it can be fit by a polynomial function with high accuracy. Regarding the periodic term, its periodicity is closely related to seasonal rainfall and regular reservoir scheduling, and thus predicting it is more complex (Du et al. 2013; Xu and Niu 2018; Yang et al. 2019).

More recently, the long short-term memory (LSTM) model has gained a lot of attention because it can efficiently reflect dynamic responses of the landslide displacement under the coupling effect of rainfall and reservoir water level (Xing et al. 2019; Zhang et al. 2019; Jiang et al. 2020). The LSTM model was first proposed by Hochreiter and Schmidhuber (1997), and can be considered as an improved method of the recurrent neural network (RNN) (Sak et al. 2014) that uses a special "memory" structure to better control and judge historical information. For instance, Fan et al. (2014) investigated the use of the LSTM model for TTS speech synthesis. Weytjens et al. (2019) found that the LSTM model is particularly useful for time-series forecasting, as compared with classical predictive methods such as ARIMA. Kong et al. (2017) efficiently used the LSTM model for predicting the short-term load forecasting problem of individual residential households.

The LSTM model has been proven to be a reliable method for building other forecasting models in many previous studies, including landslide displacement prediction problems. For instance, Zhang et al. (2019) found that the LSTM model performs better in predicting landslide displacements than the SVM and the Elman neural network (ENN) models, although it needs more time for training. Yang et al. (2019) compared the predicted displacements of the SVM and LSTM models respectively, and verified the higher accuracy of the latter. Xing et al. (2019) proposed an improved LSTM model through the stacking of neural network layers, and it has been successfully applied to the Dashuitian landslide in China. However, the performance the LSTM model on landslide displacement prediction still needs to be explored and improved through more case studies.

In this section, the LSTM model was adopted to predict landslide displacements of the Bazimen landslide in the TGR area, based on a time-series analysis and triggering factors selection. The implementation of the predictive model comprises four main tasks: (1) decomposing the total displacements into the trend term and the periodic term; (2) using a nonlinear regression function to fit the trend term; (3) building the model to predict periodic displacements by analyzing the relationships between periodic displacements and rainfall and the reservoir water level; and (4) adding the predicted trend and periodic displacements as the predicted total displacements. The SVM, MARS, and the Prophet models are also used to verify the results and compare the model performance.

6.2 Methodology: Prophet Model

The previous sections have provided the basic introduction methods of ML, only the prophet model will be basically introduced.

Prophet is an open-source algorithm developed by Facebook (Taylor and Letham 2018), and it has been used for time-series analysis based on an additive model (Zunic et al. 2020). It decomposes the time-series data into three main components: trend, seasonality, and holiday (Yan et al. 2019). The governing equation of this model can be expressed as follows (Taylor and Letham 2018):

$$y_t = g(t) + s(t) + h(t) + \varepsilon_t \tag{6.1}$$

where $g(t)$ is the trend function that reflects the long-term component in the original data, $s(t)$ denotes the periodic component which has a close relationship with time effect (e.g., yearly, monthly, weekly, and daily seasonality), $h(t)$ represents the violent changes of holidays that occur irregularly, and ε_t is the error term which is assumed to be normally distributed.

The long-term trend is the most important component of the landslide displacement, and it is mainly controlled by internal factors, e.g., geologic conditions and material properties. There are two trend models implemented in Prophet: a saturating growth model and a piecewise model. Considering similar step-wise characteristics of the landslide displacement in this work, the piecewise model was adopted to model the long-term trend, and it can be expressed as follows:

$$g(t) = \frac{C(t)}{1 + \exp(-(k + \alpha(t)^T \delta)(t - (m + a(t)^T \gamma)))} \tag{6.2}$$

where $C(t)$ is the carrying capacity of the system at any time point, $k + \alpha(t)^T \delta$ is the time-varying growth rate, δ denotes the rate adjustments, m is the offset parameters, and γ is set as correct adjustments to make the function continuous.

Due to the significant effects of external factors, such as seasonal rainfall and the periodical fluctuation of the reservoir water level, the landslide displacement usually shows a similar development pattern, that is, periodicity (Zhang et al. 2020a). To fit and forecast these effects, the seasonality function should contain periodic functions of time. Prophet uses a Fourier series which provides a flexible model. The seasonal effects $s(t)$ can be expressed as follows:

$$\mathrm{s}(t) = \sum_{n=1}^{N} (a_n \cos(\frac{2\pi nt}{P}) + b_n \sin(\frac{2\pi nt}{P})) \tag{6.3}$$

where P is a regular period which is expected as a feature of the time-series data (e.g., $P = 365$ for yearly data or $P = 7$ for weekly data) and N is the number of cycles.

For commercial purposes, the holiday effects are considered in Prophet to model and forecast unusual changes on holidays or some big events. There is no such effect on the landslide displacement, so it is not considered in this work.

6.2.1 Model Development

The landslide displacement is greatly affected by both internal and external factors, and its development can be decomposed into two components: a long-term trend and a periodic term based on specific influences. The internal factors, such as geological conditions and material properties, mainly control the landslide displacement and make it grow approximately monotonically with time at a long-time scale. The external factors mainly contain rainfall and the reservoir water level, and they have the lagging effect on the landslide displacement, which makes surficial deformation vary periodically. The accumulated displacement can be considered as a sum of these two components as follows:

$$s(t) = f(t) + g(t) \tag{6.4}$$

where $s(t)$ is the total displacement recorded by monitoring systems, $f(t)$ denotes the long-term trend displacement, $g(t)$ is the periodic displacement, and t is time.

Clearly, sudden actions (i.e., earthquakes and mine explosions) can pose a serious threat to the landslide stability and can cause rapid change within hours. However, given the existing monitoring tools and methods, the abrupt displacement caused by such random factors cannot be predicted. Due to the difficulty and uncertainty, such a scenario was not taken into consideration in this work.

6.2.2 *The Prediction Process*

Four machine learning methods were used in this work to predict the landslide displacement. The two displacement components (the trend term and the periodic term) were predicted separately via a mathematical function and machine learning methods (LSTM, MARS, and SVM), as shown in Fig. 6.1 Due to the similar mechanism of the Prophet model and the proposed model in this work, we use the former directly to predict the total displacement without any decomposition for landslide displacement, which works more efficiently.

Thus, the prediction process can be implemented as follows:

(1) The monitored total displacement should be decomposed into two components.

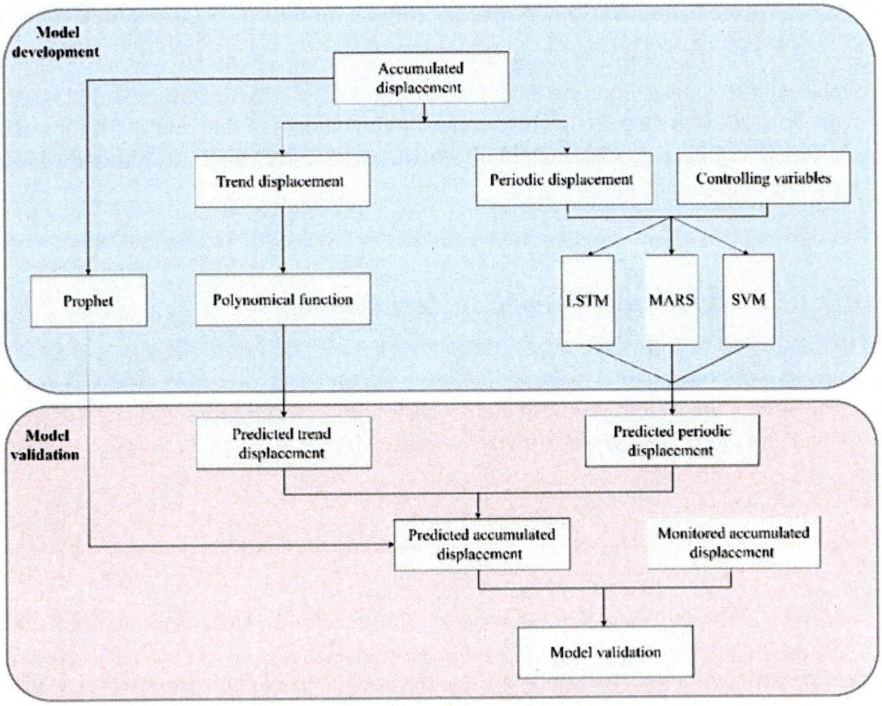

Fig. 6.1 Flowchart of the displacement prediction process

(2) The trend displacement is time-dependent, and it increases steadily over time, which can be fitted with a polynomial function with high accuracy.
(3) The periodic displacement is closely affected by rainfall and reservoir water level, and its fluctuation with time is more complex. By analyzing the correlation between rainfall, reservoir water level, and displacement, several controlling variables (seen in Sect. 6.4.3.1) were determined to predict the Bazimen landslide in the Three Gorges Reservoir area, using the mentioned methods to predict the periodic displacements, as illustrated in Fig. 6.1 In this part, the one-step forecasting method was used to predict periodic displacement at the next time step. In general, this method performs well for forecasting time-series data when prediction accuracy is required. Then, the total displacement can be obtained by adding the predicted trend and the periodic displacement.
(4) At the same stage, the total displacement can be predicted directly with the Prophet model, completely using the recorded accumulated displacement. This prediction model does not require complex data processing and relationship consideration; it is much easier to be implemented. In this case, Prophet was used to predict a multi-step displacement time series. The biggest difference from the one-step method is that this one can predict data for more time steps. This may be very helpful when decision-makers need more predicted time points to act, with a longer-term view.
(5) Finally, the predicted total displacements obtained by different machine learning methods were compared with the monitored dataset to determine the best model performance.

The entire dataset contains 175 monitoring points, which were recorded monthly from Jun. 2004 to Dec. 2018. In addition to the dataset of the year 2018 for testing, the remaining dataset was divided into two components: one for training (80%) and the other for validation (20%).

6.2.3 Evaluation of Modeling Performance

Three common statistical criteria, including the mean absolute error (MAE), the root mean square error (RMSE), and the coefficient of determination (R^2), are used here to evaluate the performance of these models. MAE is defined as follows:

$$MAE = \frac{1}{m}\sum_{i=1}^{m}|y_i - \hat{y}_i| \tag{6.5}$$

where m is the number of samples, y_i is the observed value of i_{th} sample, and $\hat{y}_i$ denotes the predicted value on the same sample i_{th}.

The RMSE is a scale-dependent measure, and it is efficient for comparing different methods applied to one same dataset. The definition of this criteria can be expressed as follows:

$$RMSE = \sqrt{\frac{1}{m}\sum_{i=1}^{m}(y_i - \hat{y}_i)^2} \quad (6.6)$$

Notations used in Eq. 6 are the same as those shown in Eq. 6.5. The lower the value of RMSE or MAE, the stronger the forecasting capacity of the model.

The R^2, defined by Eq. 6.7, is the square of the Pearson correlation coefficient (ρ) and it is usually used to evaluate the fitting quality of a model on data. It ranges from 0 to 1, with a higher value indicating a stronger correlation.

$$R^2 = \left[\frac{\sum_{i=1}^{m}(y_i - \overline{y})(\widehat{y_i} - \widetilde{y})}{\sqrt[2]{\sum_{i=1}^{m}(y_i - \overline{y})^2}\sqrt[2]{\sum_{i=1}^{m}(\widehat{y_i} - \widetilde{y})^2}}\right]^2 \quad (6.7)$$

where $\widetilde{y}$ is the average of predicted results, $\overline{y}$ denotes the mean of real values, and other notations are as same as in Eqs. 6.5 and 6.6.

6.3 Case Study

6.3.1 Background

The Bazimen landslide is located in Guizhou Village of Zigui County, on the right bank of the Xiangxi River (30°58′16″N, 110°45′30″E), which is a major tributary of the Yangtze River. It is approximately 0.8 km upstream of the estuary where the Xiangxi River meets the Yangtze River and 38 km from the Three Gorges Dam (Fig. 6.2). The slope angle ranges from 10° to 30°, with an initial sliding direction orientated at N120°. The upper boundary of the landslide is at the interface between bedrock and soil. The left and right boundaries are both defined by homologous gullies. The landslide is currently at an elevation of 139–280 m, and the original elevation of the front part was at 55 m before the first impoundment of the TGR in 2003. The landslide is fan-shaped, and the sliding body is gentle at the lower part while steep at the higher part. It covers an area of approximately 1.35×10^5 m^2, with a longitudinal dimension of 380 m and a width of 100–350 m, and has an estimated volume of 2.0×10^6 m^3.

The sliding masses of the Bazimen landslide are primarily composed of sandy shale, sandstone, gravel, and clay. Based on the borehole data, there are two main sliding surfaces found (Du et al. 2013; Yang et al. 2019). The deeper one is the initial sliding surface at a depth of 10–35 m, and the shallower one is the secondary sliding surface with a depth of 6–18 m. The materials of these two sliding surfaces are both silt and clay with fragmented rubble. The thickness of the sliding zone is thick in the middle part and thin at the upper and lower part, as shown in Fig. 6.3.

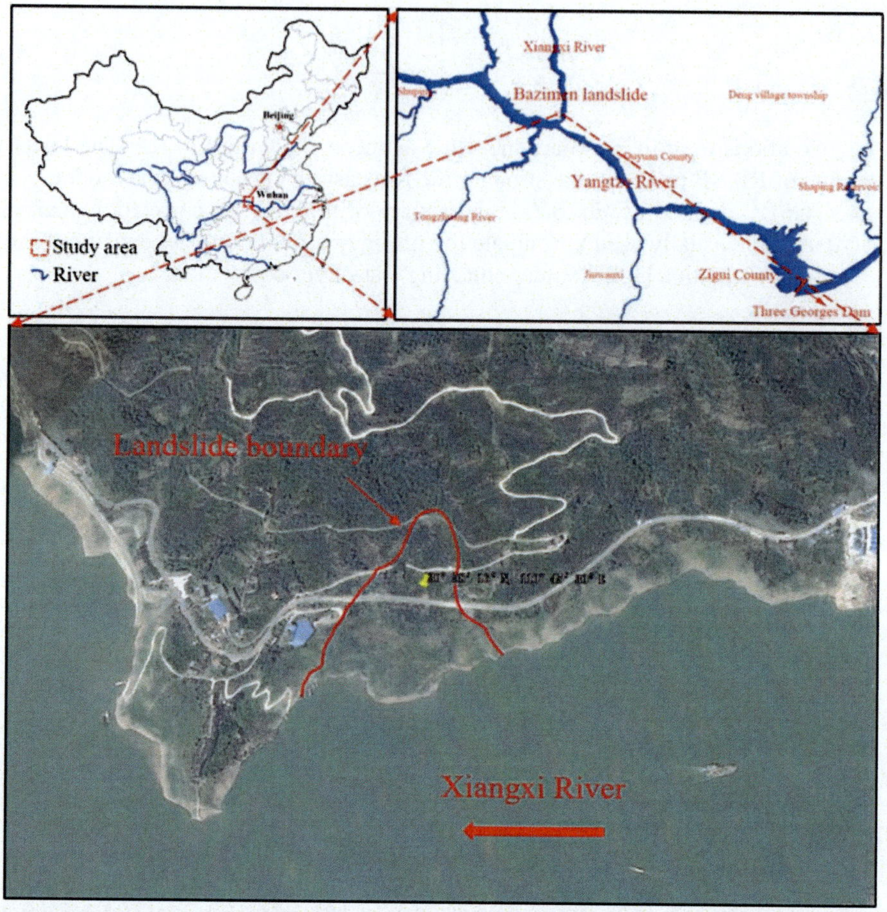

Fig. 6.2 Location and the satellite map of the Bazimen landslide in the Three Gorges Reservoir area

6.3.2 *Analysis of the Monitoring Data*

Currently, there are 10 GPS monitoring points installed on the landslide, and the numbered stations ZG110 and ZG111 have been recording data since July 2003. Figure 6.4 shows the data of the accumulated displacement, rainfall, and reservoir water level from 2003 to 2018. It can be observed that the displacement in the upper part of the landslide (monitored by ZG111) increased faster than that in the middle part (monitored by ZG110). The displacement characteristics of the landslide indicate that the development model is "advancing"; the landslide started from the upper part and developed downwards with increasing displacement. Compared with the monitored data of ZG110, the data recorded at ZG111 has more obvious piecewise characteristics, which means it can better help to explore the potential relationship

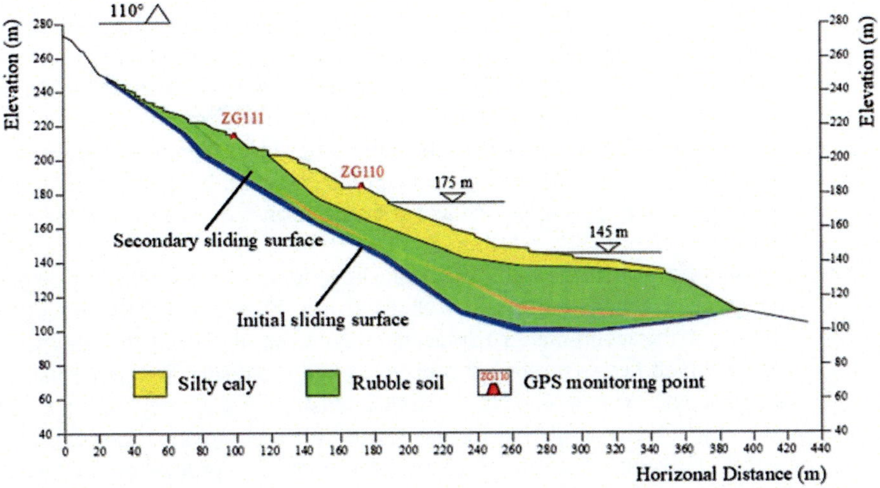

Fig. 6.3 Schematic geological cross-section I-I' of the Bazimen landslide. (modified from (Zhang et al. 2020a))

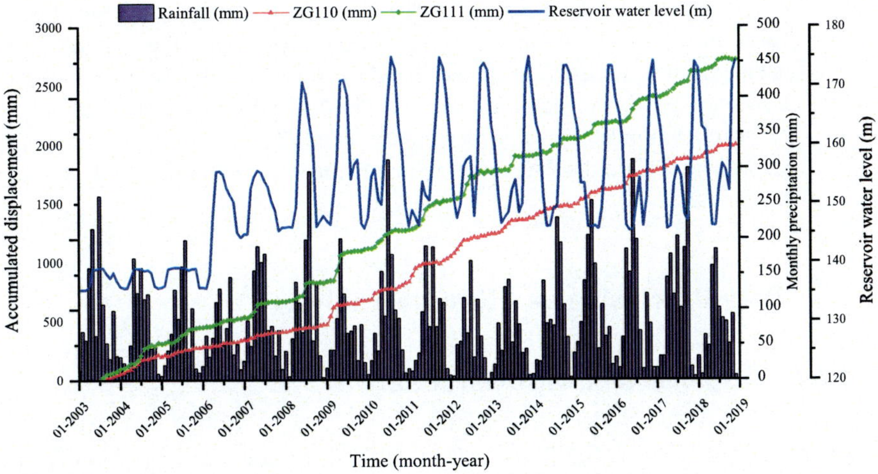

Fig. 6.4 The accumulated displacement, rainfall, and reservoir water level at Bazimen landslide

between landslide displacements and external triggering factors. It is therefore used to establish the forecasting model.

The displacement characteristics related to controlling factors can be analyzed and summarized as follows:

(1) There is a close correspondence between the scheduling of the reservoir and the accumulated displacement of the landslide: the displacement was kept almost unchanged when the reservoir water level increased or operated steadily; it

grew intensely during the period reservoir drawdown. Every time the reservoir water level dropped, the accumulated displacement rose. The reservoir water level varied annually from 2007 to 2018, and the accumulated displacement increased correspondingly within that time. However, the displacement did not respond immediately to the decline in the water level. Figure 6.4 suggests that the water level exerted a lagging effect on the accumulated displacement, and the latter always began to increase after the former decreased for one or two months.

(2) Rainfall also affects the stability of landslide, though the accumulated displacement may be weakly correlated with this factor (Zhang et al. 2020a). As shown in Fig. 6.4, the accumulated displacement grew rapidly during the rainy season, during which heavy rainfall occurs. Although the reservoir water level starts rising in late August each year, high-intensity rainfall may relent to some extent, while the landslide displacement still continues to increase in this period, indicating that rainfall plays a certain role in promoting the landslide displacement.

(3) The landslide deforms sharply under the coupling effect of rainfall and reservoir drawdown, and the monitored displacements have obvious step-wise characteristics, indicating the influence of periodic factors. Moreover, the deformation rate of the landslide increases suddenly and the accumulated displacement curve shows a steep jump during the flood season from May to August, when the maximum precipitation occurs and the reservoir water level decreases at the fastest drawdown velocity. It can be inferred that during the drawdown period of the reservoir water level, superimposed rainfall conditions will further aggravate the landslide displacement, which is more unfavorable to the stability of the landslide.

6.4 Data Processing and Results Analysis

6.4.1 Displacement Decomposition

As mentioned in Sect. 6.2.1, the accumulated displacement mainly contains two components; the long-term trend and the periodic terms. After removing the trend term from the total displacement, the residue is the periodic displacement. Many studies have shown that the moving average method is applicable for the displacement decomposition (Du et al. 2013; Yang et al. 2019), and it can be used to calculate the trend displacement as follows:

$$f(t) = \frac{y_t + y_{t-1} + \ldots + y_{t-k+1}}{k}(t = k, k + 1, \ldots, N) \tag{6.8}$$

where $f(t)$ is the trend term at time t; y_t is the accumulated displacement at time t; k is the moving average cycle, and was set as 6 months here since the rainy and dry seasons each last almost half a year; and N is the total number of samples.

6.4.2 Prediction of Trend Displacement

The trend displacement shows a stable development model with time (Fig. 6.5), and it is usually fitted with a nonlinear regression model (Xu and Niu 2018). In this work, both a linear and a cubic function were developed and compared to fit the long-term trend curve. This process was conducted using the polynomial regression model in the Scikit-learn package, backend with TensorFlow 2.0. The R^2 of a linear regression model for this dataset is 0.9956, while that of a cubic model is 0.9974, indicating that the latter one performs slightly better. The predicted values and the monitored data of the trend term are shown in Fig. 6.6, and the fitting function works well with a polynomial form, as illustrated in Eq. 6.9.

$$f(t) = -0.0003t^3 + 0.0844t^2 + 8.7555t + 174.7 \tag{6.9}$$

where $f(t)$ is the predicted value of the trend term at time t.

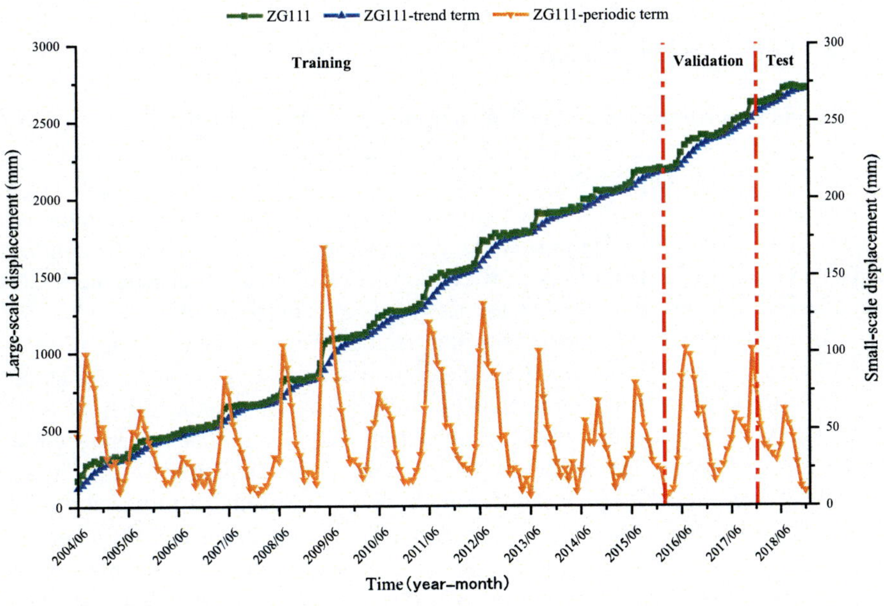

Fig. 6.5 Decomposition results of the accumulated displacement monitored at the ZG111 station

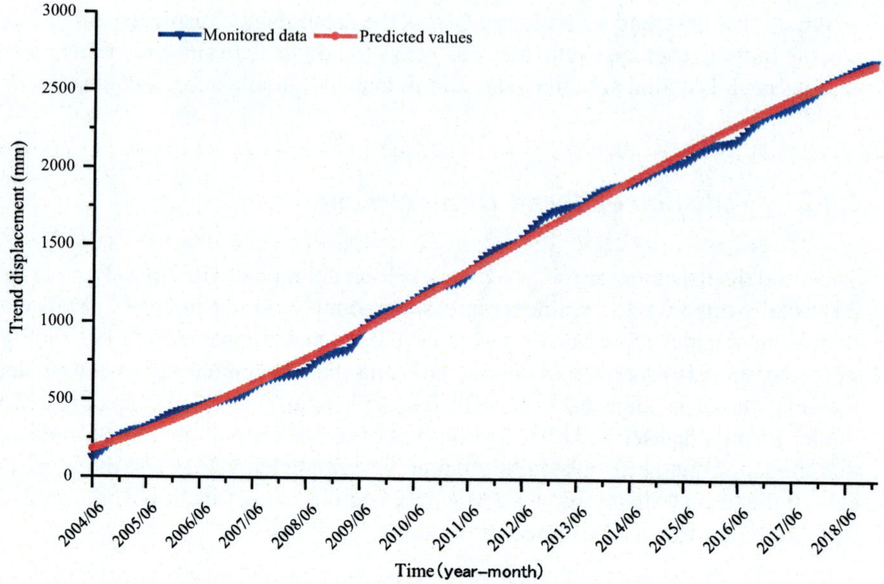

Fig. 6.6 Monitored and predicted trend displacement at ZG111 of Bazimen landslide

6.4.3 Prediction of Periodic Displacement

6.4.3.1 Controlling Factors

The model performance on the data is primarily evaluated by the accuracy, and it is thus of great importance to explore the functional relationships between the input variables and the target values. As analyzed, the coupling effect of seasonal rainfall and annual fluctuation of the reservoir water level plays an important role in triggering landslide deformation. Therefore, like common regression methods, the controlling factors (i.e., rainfall and reservoir water level) were used as input sequences and the periodic displacement was taken as the output value in this work.

In addition to the external controlling factors, the deformation state at different evolutionary stages also affects the prediction accuracy. As reported by Du et al. (2013), the performance of influencing factors depends on whether the slope is stable or not. This means even important triggering factors may not impose significant threats in terms of slope displacements when the slope is under stable conditions. In contrast, under extremely unstable conditions, even a slight disturbance can cause landslides to deform dramatically. Therefore, the stability state should be given precedence when considering influences on landslide displacement. This can be revealed by previous data, and some studies have also shown the necessity and efficiency of a concerned model when considering this factor.

The candidate variables are shown in Table 6.1, defined as v_i, where v means variable and i means the number. The Pearson's correlation coefficient (r) is also

Table 6.1 The candidate variables and the calculated Pearson's correlation coefficient

Input type	Candidate variables	r
Precipitation	v_1, the cumulative precipitation at last month	0.484
	v_2, the cumulative precipitation during last 2-month period	0.544
	v_3, the cumulative precipitation during last 3-month period	0.523
Reservoir water level	v_4, the reservoir level at current month	0.187
	v_5, the change of the reservoir level at last month	0.224
	v_6, the decline of the reservoir level at last month	0.155
Deformation state	v_7, the increased displacement over last month	0.755
	v_8, the increased displacement over last 2 months	0.626
	v_9, the increased displacement over last 3 months	0.488

calculated to measure the association strength between input variables and the periodic displacement in Table 6.1. Finally, considering the r values, v_2, v_3, v_4, v_5, v_7, and v_8 were selected as input variables to build the model.

6.4.3.2 Development for the LSTM Model

The entire dataset was divided into three parts: the data of year 2018 for testing, 80% of the remaining dataset for training, and the last 20% for validation (Fig. 6.5). Data in the training dataset were used for model construction and the validation dataset was used to verify the prediction capacity. Considering the size of this dataset, which contains 175 data points, test dataset was also used in this work to check the model's generalization ability. In some cases, when the dataset only has a small number of data points, the interpolation method may be helpful, and the test dataset can be omitted. The model constriction was implemented with TensorFlow 2.0 in the Python environment.

As aforementioned, the LSTM model is very sensitive to the data structure, and the normalization will avoid convergence problems to some extent. Therefore, both the input variables and the target values were normalized to $[-1, 1]$. The grid search method was used to optimize the model, as it can achieve a balance between the model accuracy and computational time. The process of this method can be expressed as follows: reasonable values of each parameter are combined via the cross-validation method in the training set, and then the best parameter combination can be automatically determined. In essence, it is a kind of traversing method, and it may be time-consuming to determine the optimal combination of parameters, depending on the complexity of models. Considering the scale of the dataset, a four-layer LSTM model was developed, with the first three layers as LSTM layers and the last layer as a dense layer.

The LSTM model uses previous data (selected variables as input sequences) to build potential relationships for forecasting target values, and the time step is an important parameter to be considered, which shows how many time steps of previous data are considered for forecast. The grid search method shows that the optimal time step is 3 in this work, which means the latest data are more useful to build the LSTM model. The gird search method was also used for optimizing parameters of other three machine learning methods mentioned in this work: MARS, SVM, and Prophet.

6.4.3.3 Predicted Periodic Displacement

The predicted and monitored periodic displacements at the ZG111 observation station are shown in Fig. 6.7, where the training and validation dataset using the MARS, SVM, and LSTM models are presented separately. The comparisons between the evaluation criteria of the datasets and models are illustrated in Table 6.2. It can be observed that the LSTM model performed well in both training and validation datasets. The RMSE and MAE values were 8.79 and 6.34 mm in the training dataset and were 8.72 and 6.02 mm in the validation dataset, respectively. Compared with other models, the LSTM model shows a better forecasting capacity and the results indicate that this dynamic model can capture the complex relationships between input variables and target values. In terms of the SVM model, a comparison between the results obtained by different kernel functions shows that the SVM model with a RBF kernel has better prediction capacity. Therefore, the SVM model with the RBF kernel function was used as a representative of SVM models for later analysis. Compared with the SVM model, the MARS model has a similar forecasting performance on the two datasets, with close evaluation criteria and predicted values.

In general, the displacements predicted by the MARS and SVM methods are basically consistent with the monitored values, except the local peak and irregular displacements. For example, for periodic displacements during November 2016, October 2017, and August 2018, the predicted values by the two models differ significantly from monitored values. This indicates that the prediction capacity of a static model can be seriously affected when external triggering factors (i.e., rainfall and reservoir water level) change rapidly. Although the LSTM has good forecasting capacity overall, this phenomenon is also particularly obvious for this model, which uses previous information to predict target values at the current time step. When a local peak or an irregular displacement occurs, this model still uses what it has learned from the conventional and regular data, which may cause large errors. Furthermore, the peak displacements always occur during the flood season, when the coupling effect of high-intensity rainfall and the rapid change of the reservoir water level is stronger. It is therefore notable that a model's prediction performance on irregular and unconventional data should be examined in future work.

6.4.3.4 Predicted Accumulated Displacement

As mentioned in Sect. 6.2.1, for the MARS, SVM, and LSTM models, the predicted accumulated displacement should be the sum of the predicted trend and periodic

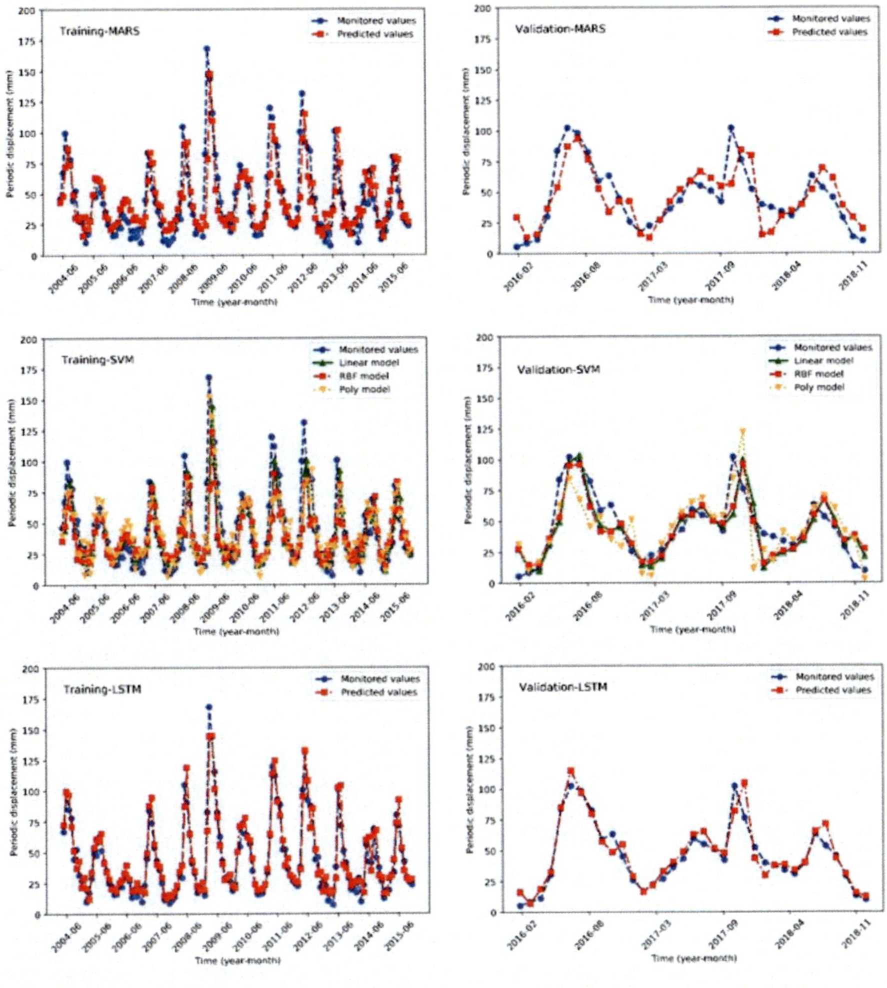

Fig. 6.7 Predicted and monitored periodic displacements at ZG111 with the MARS, SVM and LSTM models

Table 6.2 Comparison of evaluation criteria for periodic displacements with the MARS, SVM, and LSTM models

Models	Training dataset		Validation dataset	
	RMSE (mm)	MAE (mm)	RMSE (mm)	MAE (mm)
MARS	19.11	13.38	15.79	12.16
SVM (RBF)	19.73	13.47	14.40	10.76
LSTM	8.79	6.34	8.72	6.02

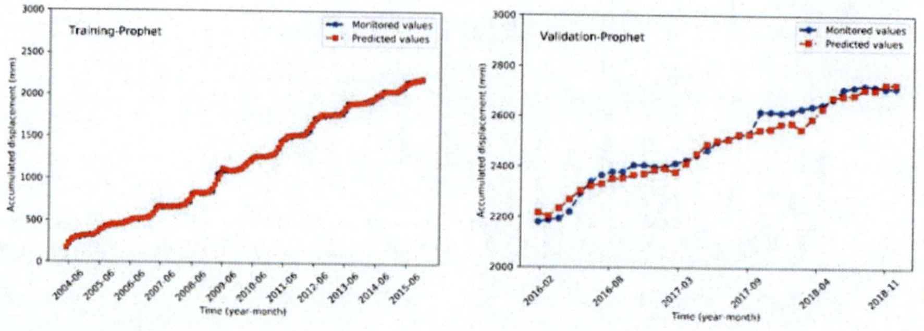

Fig. 6.8 Predicted and monitored accumulated displacements at ZG111 with the Prophet model

displacements. The trend displacement can be obtained by Eq. (6.9), which was fit based on historical displacements. For the Prophet model, due to the similarity of its mechanism and the proposed model of this work, this model was used directly to predict the total displacements based on recorded values. This model does not require the process of displacement decomposition, nor does it need previous data of a specific prior time step to predict displacements of multiple time steps.

Contrary to the other three methods that attempt to find relationships between landslide displacements and external triggering factors, the Prophet model tries to capture the time characteristics of displacements. It is thus very convenient and simple for users, as this model does not require finding candidate variables but the displacement itself. Figure 6.8 shows the Prophet model predictions for nearly 3-year displacements (February 2016 to December 2018), based on the explored model it has learned from previous data (June 2004 to January 2016). Although the predicted total displacements do not agree well with the monitored data, the step-wise development pattern was well captured by this model. In some sense, it is the real forecasting model, because it predicts based on learned development patterns rather than a simple function of input and output values, and more importantly, this model does not require extra input variables except historical displacements.

Taking the 2018 displacements as the test dataset, the predicted and monitored values shown in Fig. 6.9, and the evaluation criteria listed in Table 6.3, we arrive at the following conclusion. Generally, the LSTM model shows the highest prediction accuracy, the MARS and the SVM models have similar predicted results with medium accuracy, and the Prophet model has relatively poor prediction capacity. The values of the RMSE and MAE with the LSTM model are 18.78 and 17.03 mm respectively, and these values of the Prophet model are 36.37 and 30.09 mm.

6.5 Discussion

In this section, the LSTM model performs better both in predicting periodic displacements and accumulated displacements with Eq. 6.9. This model uses an input

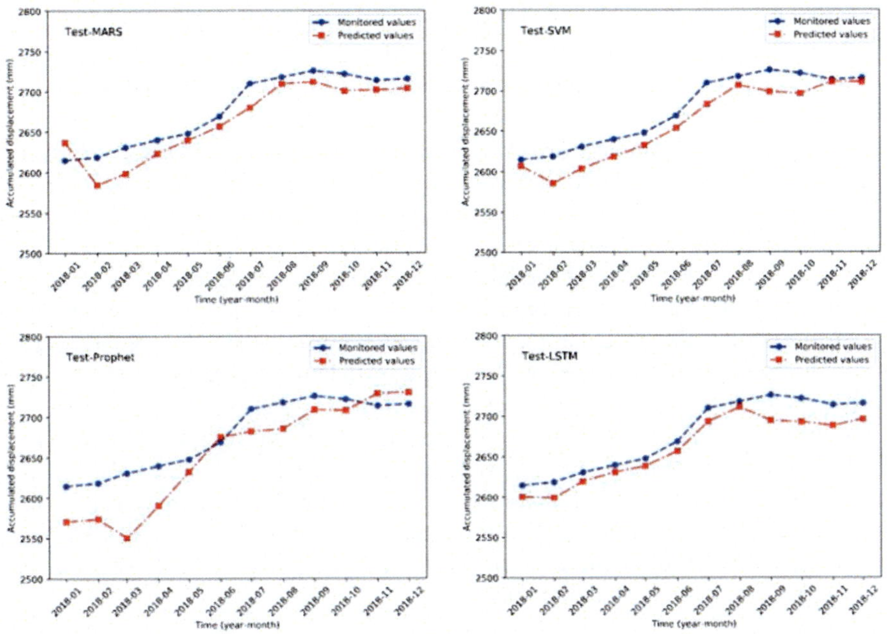

Fig. 6.9 Predicted and monitored accumulated displacements at ZG111 with the MARS, SVM, LSTM and Prophet models

Table 6.3 Comparison of evaluation criteria for accumulated displacements with the MARS, SVM, and LSTM models

Models	Test dataset	
	RMSE (mm)	MAE (mm)
MARS	20.56	18.56
SVM (RBF)	20.62	18.18
LSTM	18.78	17.03
Prophet	36.37	30.09

sequence at a specific length, containing information with a period of history, to predict target values at the current time step. Different from the other three methods mentioned in this work, this model can make full use of historical information, based on the special "gate" design. This also makes it the most time-consuming for building, developing, and testing. The LSTM model build dynamic connections between previous input variables at different time steps and target values, which means learned experience is continuously utilized and modified for current tasks. During this process, the model can update the potential rules by memorizing useful information and removing invalid rules. In other words, the memory cell of this model makes it possible to better reflect the dynamic evolution process of the landslide.

For the same datasets, the two classical machine learning methods, the SVM and the MARS models have poorer model performance. The lack of a similar memory cell in the two models may be the most important reason for this. These two models do not build connections between previous data at different time steps and the target values, which means that the models can only use historical information at a specific time point to predict the displacement correspondingly. However, the latest data structures differ from historical ones, and this may lead to errors to some extent. It is therefore suggested by some studies that in order to establish a more accurate and reliable forecasting model, the datasets should be updated timely by removing old data and adding new data (Du et al. 2013; Yang et al. 2019).

In terms of the Prophet model, its performance is inferior to the above three models. However, its advantage does not lie in the accuracy of a short forecasting period. This model implemented the time series analysis automatically and it predicted target values based on the learned trend and seasonal features. Moreover, this model does not require complex data processing and variables selection, which means it is much easier and more convenient to be carried out. In addition, because this model is based on capturing characteristics between time and displacement, it is possible to predict the target values over a longer time period than values at the next time step. It can be reasonably inferred that this model may be very applicable for engineering purposes, where cost control and time management are planned from a longer-term perspective rather than a specific time step.

6.6 Summary of this Chapter

In this section, there are four machine learning methods applied for the slope displacement prediction, including MARS, SVM, LSTM, and Prophet. The predictive results by these models generally show good consistency with the monitored displacements, which indicates that the proposed model could handle the dynamic and nonlinear relationships between the landslide displacement and external triggering factors, such as seasonal rainfall and periodic fluctuation of the reservoir water level.

Compared with the other three methods, the LSTM model shows a significant potential for dynamic landslide displacement prediction, as it builds connections between historical information and current tasks. The special design of the memory cell may be the main reason for its better performance. Given new inputs, this model can combine the memorized rules and previous information to achieve more accurate predictions. By continuously repeating this process, this model can remove invalid rules and retain useful ones to update learned relationships. In addition, this process continues throughout the entire life cycle of this model.

The SVM and MARS models show a similar forecasting ability, but their deviation is large when external triggering factors become irregular and uncommon. The static mechanisms may account for it. Compared with LSTM, both SVM and MARS try to find relationships between landslide displacement and input triggering factors at a specific time point. These two methods cannot reflect the dynamic responses hidden behind functions.

The Prophet model shows its priority and convenience in predicting displacements with a longer time period, as it can learn relationships between total displacements and time. However, it achieves the ability of predicting multi time steps at the cost of model accuracy. Therefore, it can be useful for cost control and time management problems, which require longer period information for better estimations.

References

Cao Y, Yin K, Alexander DE, Zhou C (2016) Using an extreme learning machine to predict the displacement of step-like landslides in relation to controlling factors. Landslides 13:725–736. https://doi.org/10.1007/s10346-015-0596-z

Chen L, Zhang W, Zheng Y et al (2020) Stability analysis and design charts for over-dip rock slope against bi-planar sliding. Eng Geol 275:105732

Du J, Yin K, Lacasse S (2013) Displacement prediction in colluvial landslides, Three Gorges Reservoir, China. Landslides 10:203–218. https://doi.org/10.1007/s10346-012-0326-8

Fan Y, Qian Y, Xie F-L, Soong FK (2014) TTS synthesis with bidirectional LSTM based recurrent neural networks

Hochreiter S, Schmidhuber J (1997) Long short-term memory. Neural Comput 9:1735–1780. https://doi.org/10.1162/neco.1997.9.8.1735

Huang F, Huang J, Jiang S, Zhou C (2017) Landslide displacement prediction based on multivariate chaotic model and extreme learning machine. Eng Geol 218:173–186

Jian W, Xu Q, Yang H, Wang F (2014) Mechanism and failure process of Qianjiangping landslide in the Three Gorges Reservoir, China. Environ Earth Sci 72:2999–3013

Jiang H, Li Y, Zhou C et al (2020) Landslide displacement prediction combining LSTM and SVR algorithms: a case study of Shengjibao landslide from the Three Gorges Reservoir area. Appl Sci 10:7830

Keqiang H, Xiangran L, Xueqing Y, Dong G (2008) The landslides in the Three Gorges Reservoir Region, China and the effects of water storage and rain on their stability. Environ Geol 55:55–63

Kong W, Dong ZY, Jia Y et al (2017) Short-term residential load forecasting based on LSTM recurrent neural network. IEEE Trans Smart Grid 10:841–851

Li D, Yin K, Leo C (2010) Analysis of Baishuihe landslide influenced by the effects of reservoir water and rainfall. Environ Earth Sci 60:677–687

Lian C, Zeng Z, Yao W, Tang H (2015) Multiple neural networks switched prediction for landslide displacement. Eng Geol 186:91–99

Sak H, Senior A, Beaufays F (2014) Long short-term memory recurrent neural network architectures for large scale acoustic modeling. In: Proceedings of the annual conference of the International speech communication association, INTERSPEECH, pp 338–342

Taylor SJ, Letham B (2018) Forecasting at scale. Am Stat 72:37–45

Wang F-W, Zhang Y-M, Huo Z-T et al (2004) The July 14, 2003 Qianjiangping landslide, three gorges reservoir, China. Landslides 1:157–162

Wang L, Tang L, Wang Z et al (2020a) Probabilistic characterization of the soil-water retention curve and hydraulic conductivity and its application to slope reliability analysis. Comput Geotech 121:103460

Wang L, Wu C, Tang L et al (2020b) Efficient reliability analysis of earth dam slope stability using extreme gradient boosting method. ACTA Geotech

Weytjens H, Lohmann E, Kleinsteuber M (2019) Cash flow prediction: MLP and LSTM compared to ARIMA and Prophet. Electron Commer Res 1–21

Xie P, Zhou A, Chai B (2019) The application of long short-term memory (LSTM) method on displacement prediction of multifactor-induced landslides. IEEE Access 7:54305–54311. https://doi.org/10.1109/ACCESS.2019.2912419

Xing Y, Yue J, Chen C et al (2019) Dynamic displacement forecasting of Dashuitian landslide in china using variational mode decomposition and stack long short-term memory network. Appl Sci 9:2951

Xu S, Niu R (2018) Displacement prediction of Baijiabao landslide based on empirical mode decomposition and long short-term memory neural network in Three Gorges area, China. Comput Geosci 111:87–96. https://doi.org/10.1016/j.cageo.2017.10.013

Yan J, Wang L, Song W et al (2019) A time-series classification approach based on change detection for rapid land cover mapping. ISPRS J Photogramm Remote Sens 158:249–262

Yang B, Yin K, Lacasse S, Liu Z (2019) Time series analysis and long short-term memory neural network to predict landslide displacement. Landslides 16:677–694. https://doi.org/10.1007/s10346-018-01127-x

Yin Y, Wang H, Gao Y, Li X (2010) Real-time monitoring and early warning of landslides at relocated Wushan Town, the Three Gorges Reservoir, China. Landslides 7:339–349

Zhang Y, Wang X, Tang H (2019) An improved Elman neural network with piecewise weighted gradient for time series prediction. Neurocomputing 359:199–208

Zhang W, Tang L, Li H et al (2020a) Probabilistic stability analysis of Bazimen landslide with monitored rainfall data and water level fluctuations in Three Gorges Reservoir, China. Front Struct Civ Eng 14:1247–1261. https://doi.org/10.1007/s11709-020-0655-y

Zhang W, Zhang R, Wu C et al (2020b) State-of-the-art review of soft computing applications in underground excavations. Geosci Front 11:1095–1106

Zhou C, Yin K, Cao Y, Ahmed B (2016) Application of time series analysis and PSO-SVM model in predicting the Bazimen landslide in the Three Gorges Reservoir, China. Eng Geol 204:108–120. https://doi.org/10.1016/j.enggeo.2016.02.009

Zhou C, Yin K, Cao Y et al (2018) Landslide susceptibility modeling applying machine learning methods: a case study from Longju in the Three Gorges Reservoir area, China. Comput Geosci 112:23–37

Zunic E, Korjenic K, Hodzic K, Donko D (2020) Application of Facebook's prophet algorithm for successful sales forecasting based on real-world data. arXiv Prepr arXiv:200507575s

Chapter 7
Prediction of Undrained Shear Strength Using XGBoost and RF Based on BO

7.1 Introduction

Soft sensitive clays are generally characterized by low shear strength and high compressibility and widely distributed in a near marine environment. Therefore, geotechnical design involving soft sensitive clays is rather challenging (D'Ignazio et al. 2016). Nowadays the need for the building on soft sensitive clays is increasing considering the recent tendency to infrastructure development and connectivity projects at the port. Hence, an accurate assessment of undrained shear strength by a reliable method is a critical issue for geotechnical engineers.

The undrained shear strength of soft sensitive clays can be assessed by various laboratory tests as well as in-situ tests. Most commonly, it is tested from situ, such as typically field vane shear testing (FVT) and piezocone cone penetration testing (CPTU). Generally, the in-situ test is costly and time-consuming. Besides, there is a growing consensus that there is inevitable uncertainty of correlation factors between the undrained shear strength and primarily CPTU parameters. As for laboratory tests, it requires relatively sophisticated and expensive sampling and laboratory testing techniques (D'Ignazio et al. 2016; Ching and Phoon 2018). The necessary geotechnical design parameters are not always directly measured from laboratory and in-situ tests, rather often estimated from empirical or numerical correlations that are developed from regression fitting to a prescribed dataset.

Since the pioneering work of Hansbo (1957), the transformation model was proposed based on empirical or semi-empirical regression analyses to evaluate the related design parameters of soil according to its properties (Mesri et al. 1975; Jamiolkowski et al. 1985; Kulhawy and Mayne 1990). However, the transformation model usually contains a significant amount of uncertainties from the soil properties, soil behavior, and site geology, etc. These may differ from the data source, against which the transformation models are calibrated (Kulhawy and Mayne 1990; D'Ignazio et al. 2016). Therefore, the limitation of such models needs to be recognized or else may result in bias as a direct consequence. To augment engineer's judgement, many the probabilistic transformation models based on Bayesian methods have been proposed

W. Zhang et al., *Application of Soft Computing, Machine Learning, Deep Learning and Optimizations in Geoengineering and Geoscience*,
https://doi.org/10.1007/978-981-16-6835-7_7

by scholars including (Ching and Phoon 2012, 2018; Wang and Cao 2013; Ching et al. 2014; Cao and Wang 2014; Wang and Aladejare 2016; Wang et al. 2018a, 2019),. Artificial intelligence (AI) techniques have developed rapidly in the past near years, many novel machine learning (ML) algorithms have been proposed and widely used in various fields, and may provide new perspective in this regard. This is in alignment with the International Society for Soil Mechanics and Geotechnical Engineering's (ISSMGE's) latest initiative to explore machine learning methods in geotechnical engineering (Ching and Phoon 2018).

Machine learning algorithms, such as artificial neural network (ANN) (Goh 1995; Teh et al. 1997; Yousefpour et al. 2011; Zhang et al. 2019), Bayesian network (BN) (Li et al. 2018; Koduru 2019), support vector machine (SVM) (Rodriguez-Galiano et al. 2015; Zhang et al. 2016; Zhou et al. 2016);([CSL STYLE ERROR: reference with no printed form.]), and multivariate adaptive regression splines (MARS) (Zhang and Goh 2013; Zhang et al. 2015; Liu et al. 2019), etc., have gradually become the alternative resolution for geotechnical problems. However, a single sophisticated algorithm may be not a consideration to build design soil parameters' surrogate model, since database derived from different sites have different soil properties, soil behavior and site geology. Consequently, there is a necessary need to apply ensemble methods in geotechnical engineering. Ensemble learning combines multiple algorithms that process different hypotheses to form a better hypothesis, so its predictions perform well (Nascimento et al. 2014; Xia et al. 2017a). Ensemble methods are better than single ML and other statistical methods as reported by (Nanni and Lumini 2009; Lessmann et al. 2015). Therefore, it is promising to encourage using it for further application in the geotechnics.

Ensemble methods can be broadly categorized into two methods according to their structures: bagging (parallel) and boosting (sequential). The bagging ensemble method is the combination of different learning algorithms, each of which generates an independent model in parallel (Xia et al. 2017b). As a famous bagging ensemble learning method, random forest (RF) is understandable because it can be regarded as a process of consensus decision trees making, each decision tree decides independently on a specific issue, and then the results are combined to make final decisions (Breiman 2001). Several researchers have also applied RF to solve geotechnical engineering problems, such as Zhou et al. (2016) investigated the feasibility of using RF to forecast surface movements induced by tunnel construction. Zhang et al. (2019) constructed RF and MARS models for prediction of pile drivability with numerous data. In boosting, the decision trees are built boosting such that each subsequent decision tree aims to reduce the errors of the previous decision tree. Each decision tree learns from its predecessors and updates the residual errors. Hence, the tree that grows subsequently is apt to learn from updated residuals (Friedman 2001). Extreme Gradient Boosting (XGBoost), which is an advanced supervised algorithm proposed by Chen and Guestrin (2016) under the boosting framework, which has been widely recognized in Kaggle machine learning competitions due to its advantages of high robustness and sufficient flexibility. So far, the XGBoost-based model is rarely used in geotechnical engineering.

Motivated by the benefits mentioned above, XGBoost-base and RF-based ensemble learning methods are developed for capturing the relationships between the undrained shear strength and various basic soil parameters in this paper. In order to reduce the dependence on the rule of thumb and brute-force search, the Bayesian optimization method is used to find the appropriate models hyper-parameters. To the best knowledge of the authors, this is the first time that the ensemble learning methods have adopted for building the design soil parameters' surrogate model. The remainder of this paper starts with the introduction of two efficient ensemble learning algorithms and Bayesian hyper-parameter optimization method, followed by introduction of the datasets investigated and its preliminary analysis in this study. Then, four performance measures were employed to analyze and assess the predictive results. The last part gives the developed models are used as a contrastive analysis with three comparison machine learning methods and two transformation models under fivefold CV for better persuasive results.

7.2 Methodology

For RF and XGBoost, the previous chapters have provided the detailed information, this section mainly focuses on methodology introduction of optimizations.

7.2.1 Bayesian Hyper-Parameter Optimization

In machine learning, hyper-parameter is parameter whose value need to be pre-set before the learning process, and algorithms are rarely hyper-parameter free. For XGBoost, RF and comparative ML algorithms, all have several hyper-parameters that have huge impact on the predictive accuracy of models. Therefore, reasonable tuning of these hyper-parameters, i.e., hyper-parameter optimization, is important. However, hyper-parameter optimization is a combinatorial optimization problem, which cannot be optimized by gradient descent method like general parameters. In addition, due to individual adjustment of the hyper-parameter requires retraining to evaluate the effect, computation for evaluating a set of hyper-parameter configurations is very demanding.

Grid search (GS) and random search (RS) methods are commonly used for tuning hyper-parameters in the machine learning practice. GS is a way that searches a suitable set of optimal hyper-parameter configurations by trying combinations of all the hyper-parameter. However, GS is apt to suffering from the dimensionality constraint, i.e., the number of times required to evaluate the model during hyper-parameter optimization grows exponentially with consideration of more parameters, which has been proved to be infeasible to XGBoost for substantial number of hyper-parameters. Additionally, it is not even guaranteed to find the optimal solution, often aliasing over the optimal configuration. RS is a way that selects a suitable set of optimal hyper-parameter configurations by trying random combinations of hyper-parameters. The drawback of RS lies in the unnecessarily high variance. The method is, entirely random, and uses no intelligence in selecting the trial points. Neither GS nor RS utilizes correlations between different combinations of hyper-parameters.

More and more hyper-parametric tuning processes are now done through automated methods that aim to find optimal hyper-parameters in less time using a strategy-based informed search. In addition, no additional manual operations are required except for the initial setup. Bayesian optimization is the top choice for optimizing objective functions (Snoek et al. 2012; Ghahramani 2015; Xia et al. 2017b). Bayesian optimization finds the value that minimizes the objective function by building a surrogate reconstruction (probability model) based on the past evaluation results of the target. It has been widely used for machine learning hyper-parameter tuning recently. The results show that the method is more generalized on the testing set and requires fewer iterations than GS and RS (Bergstra et al. 2013).

7.2.2 Sequential Model-Based Optimization (SMBO)

Sequential model-based optimization (SMBO) is a succinct formalism of Bayesian optimization and useful when discussing variations (Bergstra et al. 2011; Hutter et al. 2011; Hoffman and Shahriari 2014). In this subsection, Sequential model-based optimization will be introduced briefly. Firstly, the formalism SMBO constructs a model S_0 to map the hyper-parameter settings λ to the loss function L (line 1 in Table 7.1), which is used to evaluate a specific hyper-parameter setting. Then the settings λ and the corresponding loss are stored in an H (or trails), which assists in the inspection inside the Bayesian hyper-parameter optimization and displays the parameter settings and the corresponding evaluations. Afterward, SMBO iterates the following four steps: find the local optimal hyper-parameter settings λ^* on the basis of the current model S_{t-1}; calculate the loss c under the settings λ^*; update H with storage of λ^* and the corresponding loss c; and construct a new model based on the updated H. The SMBO outputs the global optimal hyper-parameter settings with the

Table 7.1 Sequential model-based global optimization (SMBO)

1: *Initialization* S_0; $H = \emptyset$
2: **for** $t = 1$ *to* T **do**
3: $\lambda^* = \text{argmin} S_{t-1}(\lambda)$
4: $c = L(\lambda^*)$
5: $H = H \cup (\lambda^*, c)$
6: *fit new model* S_t*according to updated H*
7: **end for**
8: **Return** λ *with minimum c in H*

minimum c until reaches the pre-defined maximum iteration T, which indicates the end of the iteration loop.

Tree Parzen Estimators (TPE).

Hyper-parameter optimization is fundamentally a problem of optimizing a specific mapping function over graph-structured configuration space. In order to operate in the standard SMBO algorithm, various probabilistic regression models can be used in the Bayesian optimization context, such as Gaussian Processes, Random Forests and Tree Parzen Estimators (TPE) et al.

As suggested by Bergstra et al. (2011), the configuration space is restricted to a tree structure in TPE which provides simple solutions to determine the model S (i.e., line 6 in Table 7.1) and find the local optimal hyper-parameter settings (i.e., line 3 in Table 7.1). This model does not define a predictive distribution over the objective function; instead, it creates two hierarchical processes, $l(\lambda)$ and $g(\lambda)$ acting as generative models for all domain variables. These processes model the domain variables when the objective function is below and above a specified quantile c^*,

$$p_S(c|\lambda) = \begin{cases} l(\lambda), & c < c^* \\ g(\lambda), & c \geq c^* \end{cases} \tag{7.1}$$

where $l(\lambda)$ is the density estimate constructed by the observations regarding the value of the loss, which is lower than c^* in H, whereas $g(\lambda)$ is generated by the remaining observations.

Simultaneously, a more important issue is emerged to determine a local optimal hyper-parameter setting λ^* according to the current model S_t for the t-times loop. SMBO adopts acquisition function to process this step that gathers the predictive information of S_t at any possible hyper-parameter settings till it reaches its maximum, the most promising settings are then acquired. Two forms of acquisition function are widely adopted, namely expected improvement (EI) and the probability of improvement. In this paper, EI is chosen as the criterion due to its satisfactory performance and intuitive (Bergstra et al. 2011). The improvement can be expressed as:

$$\mathrm{I}(\lambda) = \max(c_{\min} - c(\lambda), 0) \tag{7.2}$$

Actually, the improvement is difficult to estimate directly, because $c(\lambda)$ is unknown until the settings are evaluated by a loss function (i.e., line 4 in Table 6.4). Nevertheless, it is worth noting that EI can be computed as:

$$\mathrm{EI}(\lambda) = \int_{-\infty}^{c_{\min}} \max(c_{\min} - c(\lambda), 0) \cdot p_{S_t}(c|\lambda)\mathrm{d}c \tag{7.3}$$

Among which, $c_{\min}$ denotes the current minimum loss in H, and $c(\lambda)$ represents the loss under the hyper-parameter setting λ.Through all the above processes, the hyper-parameters of aforementioned two ensemble learning algorithms and three comparison algorithms all can be selected by Bayesian optimization and some details of the parameter adjustment process can be referred to Bergstra et al. (2013).

7.3 Databases

This study employed the datasets of F-CLAY/7/216 and S-CLAY/7/168 (D'Ignazio et al. 2016), which are available in the TC304 database. The first one F-CLAY/7/216 was compiled from 216 samples of field vane tests, which was conducted in 24 different test sites in Finland. The second clay dataset S-CLAY/7/162 consisted of 168 data points of field vane tests from 12 sites in Sweden and 7 sites in Norway. Each data points in the two datasets include six parameters, including undrained shear strength (USS), vertical effective stress (VES), preconsolidation stress (PS), liquid limit (LL), plastic limit (PL) and natural water content (W). The two datasets were combined and studied together herein. Figure 7.1 shows the histograms of undrained shear strength (USS) and relevant input features as well as their mean and standard deviation, and the eight parameters their distribution approximates a normal distribution or a lognormal distribution. Table 7.2 lists the statistics with the mean value, the maximum (Max), minimum (Min), coefficient of variation (COV), and the sample numbers (n). It should be noted that VES and PS have the largest variations.

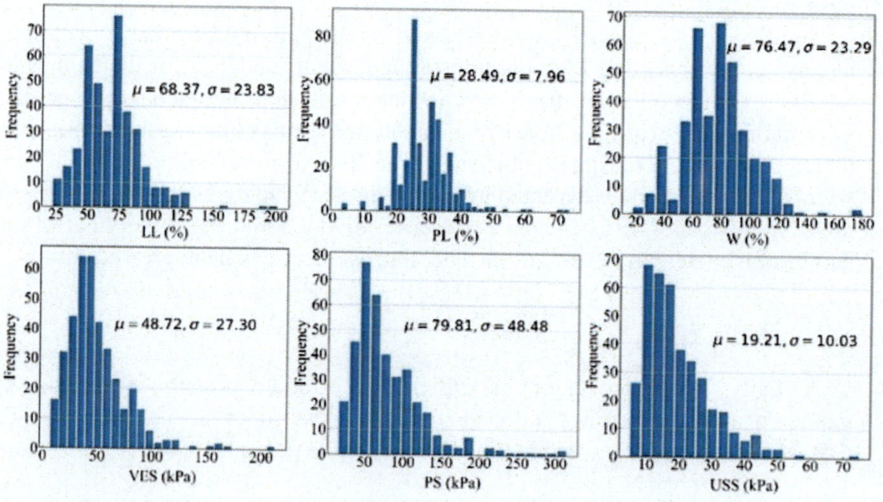

Fig. 7.1 Histogram of the frequency distribution of features and label

Table 7.2 Basic statistics of the five feature variables and label

	LL (%)	PL (%)	W (%)	VES (kPa)	PS (kPa)	USS (kPa)
mean	68.37	28.49	76.47	48.72	79.82	19.21
cov	0.35	0.28	0.31	0.56	0.61	0.52
max	201.81	73.92	180.11	212.87	315.64	75.00
min	22.00	2.73	17.27	6.86	15.20	5.00
n	384	384	384	384	384	384

Table 7.3 Correlation partition based on r_s values

r_s	Level	Parameters
0–0.19	Very weak	USS versus LL, USS versus PL
0.2–0.39	Weak	USS versus W, PS versus LL, PS versus PL, VES versus LL, VES versus PL
0.4–0.59	Moderate	USS versus VES, PS versus W, VES versus W
0.6–0.79	Strong	USS versus PS, PS versus VES, W versus PL, PL versus LL
0.80–1.00	Very strong	W versus LL

7.3.1 Correlation Analysis

Generally, the correlation coefficient of multi-source data can be displayed together in the heat map, as the correlations between all parameters can be clearly and concisely drawn. Based on the data distribution, the Spearman's rank correlation coefficient (r_s) was applied to determine the correlations between each pairwise variables, then the heat map consisting all correlation coefficients were formed. Table 7.3 clearly summarized the correlation of all parameters according to the r_s (absolute value) falling at different intervals. As shown in Fig. 7.2, it can be easily figured out that the PS is strongly correlated with USS ($r_s = 0.75$), meanwhile the VES is moderate correlated with USS ($r_s = 0.53$), while USS is almost independent from the LL and PL ($|r_s| \leq 0.11$).

7.3.2 Removal of Outliers for USS

The data cleaning process was promoted to filter out some outliers in USS data, since these very small or large values may bring bias to the developed model. As can be observed from the histogram, the distribution of input variables did not strictly normal distributed. Therefore, the three-sigma rule cannot properly reflect the data structure. The more general choice is the box plot, it is a way for graphically depicting databases by their quartiles. It contains the maximum, minimum, median, as well as the upper quartiles and lower quartiles for a set of data. The maximum value was

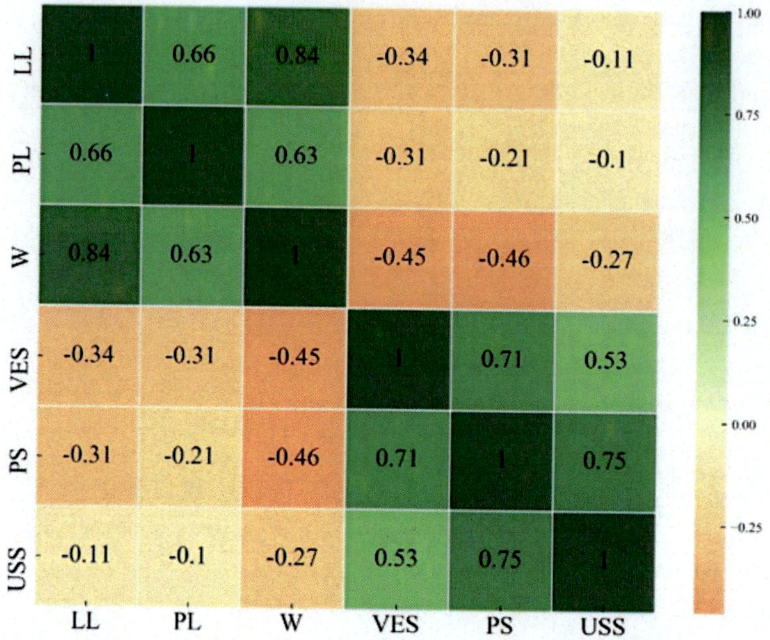

Fig. 7.2 Correlation coefficient matrix heatmap of the feature variables and the label

defined as min{3rd Quartile + 1.5*interquartile range, the actual maximum value in datasets}. Similarly, the minimum value was defined as max{1st Quartile-1.5 interquartile range, the actual minimum value in datasets}(the interquartile range is upper quartiles minus lower quartiles). As shown in Fig. 7.3, calculated by the above

Fig. 7.3 The box plot of USS

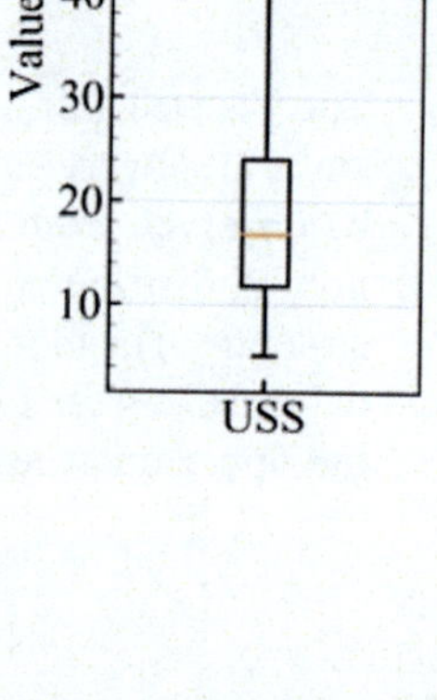

principle, the maximum value (42.34) and minimum value (5.00) of the box plot were determined. In this way, a total of 12 data points were deleted.

7.4 Implementation Procedure

7.4.1 K-fold CV

Ideally, if we have sufficient data at hand, the best practice is to randomly divide the dataset into three parts: a training set, a validation set, and a testing set. The training set is used to fit the models; the validation set is adopted to estimate prediction error for model selection; while the testing set is utilized for assessment of the generalization error of the finalized model (Franklin 2005). Since data are generally scarce (only 336 data points herein), the inability to truly reflect the generalization performance of the model is common. To avoid bias in data selection, one of the most popular validation methods, such as k-fold CV (Kohavi and Kohavi 1995; Rodríguez et al. 2010; Wong 2015; Zhou et al. 2019), was employed in this paper during the process of hyper-parameters tuning and model assessment. In k-fold CV, original samples D are randomly divided into k mutual exclusion subsets of similar size, i.e. $D = D_1 \cup D_2 \cup \ldots \cup D_k$, $D_i \cap D_j = \emptyset(i \neq j)$.Each subset D_i maintains the consistency of the data distribution as much as possible, that is, from hierarchical sampling by D. Then, each time the union of $k1$ subsets is used as the training set, and the remaining subset is adopted as the test set; thus, the k group training/testing set can be obtained, so that k training and testing can be performed. Although a definite strict rule for determining the value of k does not exist, a value of $k = 5$ (or 10) is very common in the field of applied machine learning. In this regard, the number of folds' k was set to five as suggested by Kohavi and Kohavi (1995); Wong (2015) and associated with the trade-off between the computation time and the bias. Therefore, this paper adopts the method of fivefold CV to verify the model (see Fig. 7.4).

7.4.2 Comparison Models

To evaluate the performance of the proposed XGBoost-based and RF-based design soil parameters' surrogate models, it is compared with three comparison ML algorithms, including multivariate adaptive regression splines (MARS) (Friedman 2007), support vector machine (SVM) (Cortes and Vapnik 1995) and multilayer perceptron (MLP, a class of feedforward artificial neural network) (Gardner and Dorling 1998). To be fair, all comparison ML models here also will use Bayesian optimization method for tuning hyper-parameters. Besides, two relatively good transformation models presented in D'Ignazio et al. (2016), Eqs. (7.4) and (7.5) also as a baseline model compares with above models.

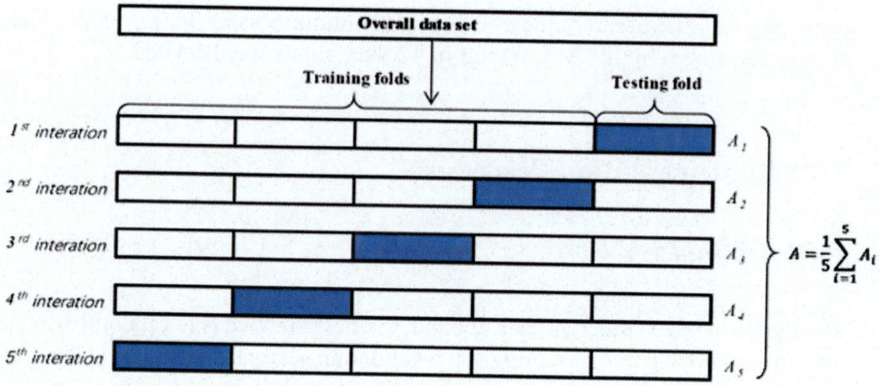

Fig. 7.4 Schematic diagram of XGBoost and RF modeling

$$\text{USS} = 0.296\text{OCR}^{0.788} W^{0.337}\text{VES} \tag{7.4}$$

$$\text{USS} = 0.319\text{OCR}^{0.757} LL^{0.333}\text{VES} \tag{7.5}$$

7.4.3 Performance Measures

The following performance measures were employed to analyze and assess the predictive results of the machine learning models: RMSE, R^2, MAPE and Bias factor *b*.

7.5 Calculation Results

In this section, the developed models were systematically compared with three comparison machine learning methods and two transformation models with respect to prediction accuracy and robustness under fivefold CV. For interest, the SVM, MLP and RF algorithms were performed using the Sklearn library. Meanwhile, the XGBoost and MARS are operated with XGBoost package and Earth package on Python 3.6. Bayesian hyper-parameter optimization is performed using the hyperopt package on Python. All the calculations are performed on a desktop computer with 3.0 GHz Intel i5 CPU, 8 GB RAM, and Windows 10 operating system.

7.5.1 Predictive Comparisons Among Different Models

Figures 7.5 and 7.6 show the RMSE curves of the training and testing sets of ML models under fivefold CV, respectively. It is well recognized that RMSE value closer to 0 indicates that the error in prediction is marginal. It can be observed that the two ensemble learning models' RMSE are smaller than the other three ML models' RMSE, indicating that XGBoost-based and RF-based models' prediction performance is better than others in training set. The curves are more fluctuant in Fig. 7.6. In addition, all models performed the same trend and get the worst at $k = 2$ and best at $k = 3$, indicating that models are obviously influenced by the data set's quality. In addition, it can be noticed that XGBoost-based model's predictive performance is obviously better than the others, while the advantages of RF-based model cannot be clearly observed. By calculating the RMSE average values for these models,

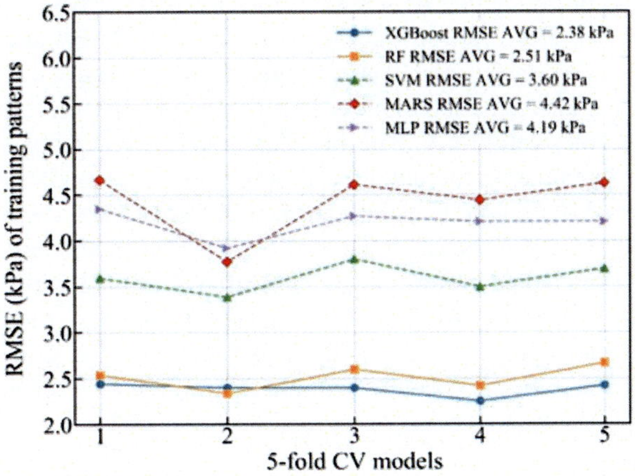

Fig. 7.5 The RMSE curves of training patterns under fivefold CV

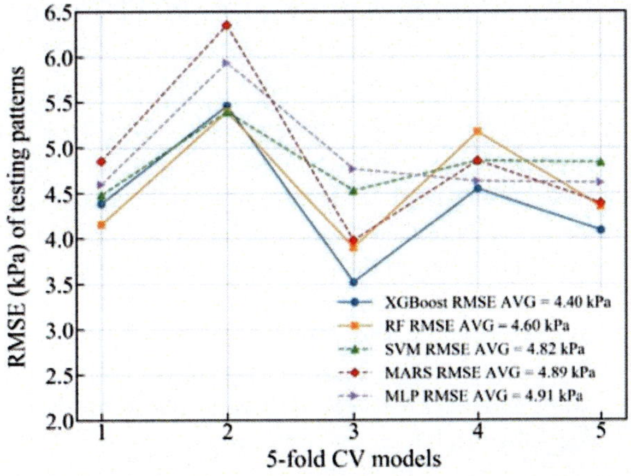

Fig. 7.6 The RMSE curves of testing patterns under fivefold CV

Table 7.4 Comparisons among models predictive modeling results

Evaluation index	RMSE (kPa)		R^2		MAPE (%)		Bias	
	Training	Testing	Training	Testing	Training (%)	Testing (%)	Training	Testing
XGBoost	2.38	4.40	0.92	0.73	10.85	19.23	0.99	1.01
RFR	2.51	4.60	0.91	0.70	10.93	19.63	0.99	1.00
SVR	3.60	4.82	0.82	0.67	13.69	20.56	1.02	1.04
MLPR	4.19	4.91	0.75	0.66	19.41	21.38	1.00	1.02
MARS	4.42	4.89	0.73	0.66	20.23	22.43	1.01	1.04
Equation (7.4)	5.88		0.52		23.39		1.07	
Equation (7.5)	6.11		0.48		23.60			

RF-based model's RMSE average value is lower than that of comparison models, implying that RF-based model is much better.

The average values of RMSE, R^2, MAPE and bias factor b under the fivefold CV for all models are summarized in Table 7.4. It should be stressed that the two transformation models, Eqs. (7.4) and (7.5), directly used the entire database instead of fivefold CV. The bias factor b derived from all models above are close to 1, indicating the unbiased property of these models. The mean values of RMSE, R^2, MAPE for the testing set of the XGBoost-based model is 4.40 kPa, 0.73 and 19.23%, respectively, and it outperformed the other ML-based models and two transformation models. Furthermore, the overall order of prediction accuracy followed by XGBoost, RF, SVM, MLP, MARS, Eqs. (7.4) and (7.5). In a word, the ensemble models have advantages in prediction accuracy over the single sophisticated algorithm in the case of multi-source data. In addition, it also outperformed the models presented in (D'Ignazio et al. 2016).

7.5.2 Fitting Performance of XGBoost and RF

Figures 7.7 and 7.8 show the fitting performance of XGBoost-based and RF-based model at $k = 3$. The fitting results of $k = 3$ are selected as a representative to show fitting performance since this set of testing data is the closest to overall data's distribution. The cyan line represents the actual USS values with sample number, and the upper boundary of the orange area is 1.2 times the actual value of USS and the lower boundary of the orange area is 0.8 times the actual value of USS. Furthermore, the blue square and red circle represent the prediction values of XGBoost-based model and RF-based model, respectively. It is apparent that most of the data points fall into the orange area, and the MAPE values of the two models for testing set are 16.92% and 18.19%, respectively. These MAPE values can be regarded as small in view that such data set contains multi-source and certain noisy data.

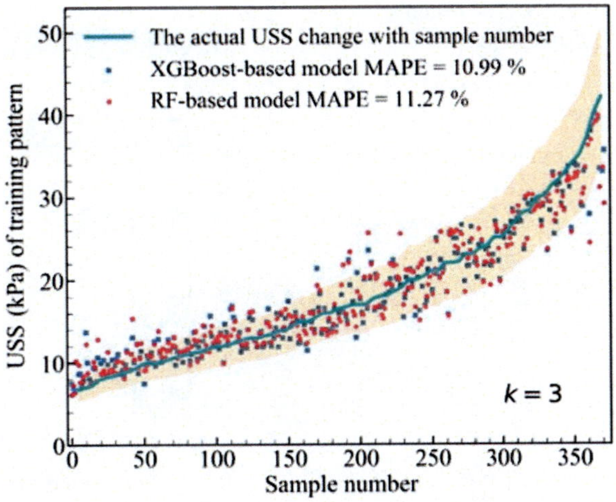

Fig. 7.7 Prediction performance of ensamble learning training pattern ($k = 3$)

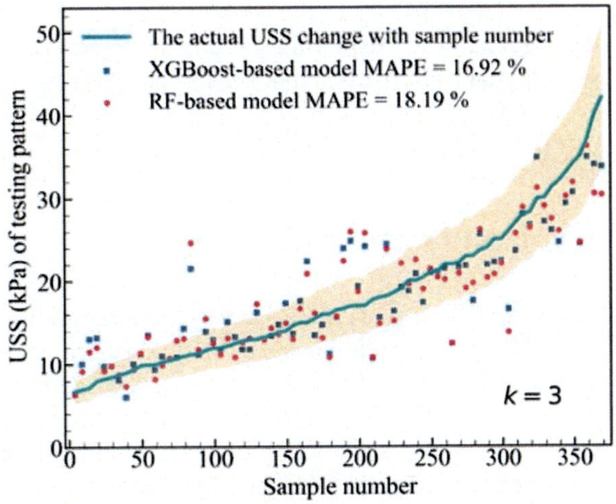

Fig. 7.8 Prediction performance of ensamble learning testing pattern ($k = 3$)

7.5.3 Features Importance Analysis

The feature importance is an important reference for feature selection and model interpretability. A trained XGBoost-based model can automatically calculate feature importance, which can be obtained through the interface feature importance criterion, i.e. gain criterion. The gain is calculated by taking each feature's contribution for each tree in the model, indicating the relative contribution of each feature to the model. A higher value of this metric, compared to another, implies greater importance of such feature for generating a prediction. Figure 7.9 shows the five feature variables' average of feature relative importance (%) under fivefold CV. In the current XGBoost-based model, PS (54.3%) is the most important feature variables, followed

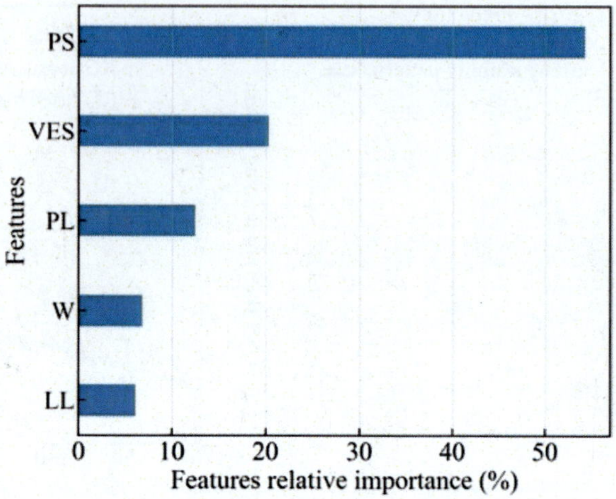

Fig. 7.9 The features relative importance ranking

by VES (20.3%), PL (12.5%), W (6.8%) and LL (6.1%). This result implies significant guidance for exploring the characteristics of the undrained shear strength of soft clays and similar result of sensitivity analyses was also presented in (D'Ignazio et al. 2016).

7.6 Summary and Conclusions

Ensemble learning methods are very popular and powerful machine learning tools for multivariate regression and classification problems. Unlike the extensive study of transformation model based on empirical evidence, data-driven ensemble learning method have received limited research attention in geotechnical engineering. This paper initiates the use of XGBoost-based and RF-based ensemble learning methods, for prediction of the undrained shear strength of soft clays. The performance of ensemble learning methods is also compared with three baseline ML algorithms (i.e., SVM, MLP and MARS) and two transformation models from previous works, in a combination database from F-CLAY/7/216 and S-CLAY/7/168. Based on this study, the following conclusions may be drawn.

All proposed ML models are validated over database using four performance measures under fivefold CV. The comparisons with baseline models show the superiority of the XGBoost-based and RF-based model in terms of predictive performance, and the overall order of prediction accuracy followed by XGBoost RF, SVM, MLP, MARS.

ML methods have obvious advantages in prediction accuracy compared with previous transformation models. The mean MAPE value of the two ensemble learning models for testing set is 19.23% and 19.63%, respectively, and these MAPE values can be considered small for such data set containing multi-source and certain noisy data.

A Bayesian hyper-parameter optimization method, is employed to tune ML algorithms' hyper-parameters. This measure ensures that the proposed model can fully tap the predictive potential, and perform better than previous transformation models for capturing the relationships between the USS and various basic soil parameters.

The proposed XGBoost-based model can automatically calculate the relative feature importance of soil parameters in USS, PS is identified as the most important feature variables, followed by VES, PL, W and LL. This measure enhances the model's interpretative and comprehensive and provides guidance on feature selection for Geotechnical practitioners in practical applications.

References

Bergstra J, Bardenet R, Bengio Y, Kégl B (2011) Algorithms for hyper-parameter optimization. In: Advances in neural information processing systems 24: 25th annual conference on neural information processing systems 2011, NIPS 2011

Bergstra J, Yamins D, Cox D (2013) Hyperopt: a python library for optimizing the hyperparameters of machine learning algorithms. In: Proceedings of the 12th python in science conference. SciPy, pp 13–19

Breiman L (2001) Random forests. Mach Learn 45:5–32. https://doi.org/10.1023/A:1010933404324

Cao Z, Wang Y (2014) Bayesian model comparison and characterization of undrained shear strength. J Geotech Geoenviron Eng 140:04014018. https://doi.org/10.1061/(asce)gt.1943-5606.0001108

Chen T, Guestrin C (2016) XGBoost: a scalable tree boosting system. In: Proceedings of the ACM SIGKDD international conference on knowledge discovery and data mining. Association for Computing Machinery, pp 785–794

Ching J, Phoon K-K (2018) Constructing site-specific multivariate probability distribution model using Bayesian machine learning. J Eng Mech 145:04018126. https://doi.org/10.1061/(ASCE)EM.1943-7889.0001537

Ching J, Phoon KK (2012) Modeling parameters of structured clays as a multivariate normal distribution. Can Geotech J 49:522–545. https://doi.org/10.1139/T2012-015

Ching J, Phoon KK, Chen CH (2014) Modeling piezocone cone penetration (CPTU) parameters of clays as a multivariate normal distribution. Can Geotech J 51:77–91. https://doi.org/10.1139/cgj-2012-0259

Cortes C, Vapnik V (1995) Support-vector networks. Mach Learn 20:273–297. https://doi.org/10.1023/A:1022627411411

D'Ignazio M, Phoon KK, Tan SA, Länsivaara TT (2016) Correlations for undrained shear strength of finnish soft clays. Can Geotech J 53:1628–1645. https://doi.org/10.1139/cgj-2016-0037

Franklin J (2005) The elements of statistical learning: data mining, inference and prediction. Math Intell 27:83–85. https://doi.org/10.1007/bf02985802

Friedman JH (2001) Greedy function approximation: a gradient boosting machine. Ann Stat 29:1189–1232. https://doi.org/10.1214/aos/1013203451

Friedman JH (2007) Multivariate Adaptive Regression Splines. Ann Stat 19:. https://doi.org/10.1214/AOS/1176347963

Gardner MW, Dorling SR (1998) Artificial neural networks (the multilayer perceptron)—a review of applications in the atmospheric sciences. Atmos Environ 32:2627–2636. https://doi.org/10.1016/S1352-2310(97)00447-0

Ghahramani Z (2015) Probabilistic machine learning and artificial intelligence. Nature 521:452–459

Goh ATC (1995) Empirical design in geotechnics using neural networks. Geotechnique 45:709–714. https://doi.org/10.1680/geot.1995.45.4.709

Hansbo S (1957) A new approach to the determination of the shear strength of clay by the fall-cone test. R Swedish Geotech Inst Proc 46

Hoffman MW, Shahriari R (2014) Modular mechanisms for Bayesian optimization. NIPS Work Bayesian Optim 1–5

Hutter F, Hoos HH, Leyton-Brown K (2011) Sequential model-based optimization for general algorithm configuration. In: Lecture notes in computer science (including subseries lecture notes in artificial intelligence and lecture notes in bioinformatics). Pp 507–523

Jamiolkowski M, Ladd CC, Germaine JT, Lancellotta R (1985) New developments in field and laboratory testing of soils. Proc 11th Int Conf soil Mech Found Eng San Fr August 1:57–153

Koduru S (2019) A bayesian network for slope geohazard management of buried energy pipelines. In: 13th International conference on applications of statistics and probability in civil engineering, ICASP 2019

Kohavi R, Kohavi R (1995) A study of cross-validation and bootstrap for accuracy estimation and model selection

Kulhawy FH, Mayne PW (1990) Manual on estimating soil properties for foundation design. Ostigov 299

Lessmann S, Baesens B, Seow HV, Thomas LC (2015) Benchmarking state-of-the-art classification algorithms for credit scoring: an update of research. Eur J Oper Res 247:124–136. https://doi.org/10.1016/J.EJOR.2015.05.030

Li X, Zhang L, Zhang S (2018) Efficient Bayesian networks for slope safety evaluation with large quantity monitoring information. Geosci Front 9:1679–1687. https://doi.org/10.1016/J.GSF.2017.09.009

Liu L, Zhang S, Cheng YM, Liang L (2019) Advanced reliability analysis of slopes in spatially variable soils using multivariate adaptive regression splines. Geosci Front 10:671–682. https://doi.org/10.1016/j.gsf.2018.03.013

Manual on estimating soil properties for foundation design (Technical Report) | OSTI.GOV. https://www.osti.gov/biblio/6653074. Accessed 20 Aug 2021

Mesri G, Rokhsar A, Bohor BF (1975) Composition and compressibility of typical samples of Mexico City Clay. Geotechnique 25:527–554. https://doi.org/10.1680/GEOT.1975.25.3.527

Nanni L, Lumini A (2009) An experimental comparison of ensemble of classifiers for bankruptcy prediction and credit scoring. Expert Syst Appl 36:3028–3033. https://doi.org/10.1016/j.eswa.2008.01.018

Nascimento DSC, Coelho ALV, Canuto AMP (2014) Integrating complementary techniques for promoting diversity in classifier ensembles: a systematic study. Neurocomputing 138:347–357. https://doi.org/10.1016/j.neucom.2014.01.027

Rodriguez-Galiano V, Sanchez-Castillo M, Chica-Olmo M, Chica-Rivas M (2015) Machine learning predictive models for mineral prospectivity: an evaluation of neural networks, random forest, regression trees and support vector machines. Ore Geol Rev 71:804–818. https://doi.org/10.1016/J.OREGEOREV.2015.01.001

Rodríguez JD, Pérez A, Lozano JA (2010) Sensitivity analysis of k-fold cross validation in prediction error estimation. IEEE Trans Pattern Anal Mach Intell 32:569–575. https://doi.org/10.1109/TPAMI.2009.187

Snoek J, Larochelle H, Adams RP (2012) Practical Bayesian optimization of machine learning algorithms BT—Proceedings of the 25th international conference on neural information processing systems 2:2951–2959

Teh CI, Wong KS, Goh ATC, Jaritngam S (1997) Prediction of pile capacity using neural networks. J Comput Civ Eng 11:129–138. https://doi.org/10.1061/(asce)0887-3801(1997)11:2(129)

Wang L, Cao ZJ, Li DQ et al (2018) Determination of site-specific soil-water characteristic curve from a limited number of test data – A Bayesian perspective. Geosci Front 9:1665–1677. https://doi.org/10.1016/j.gsf.2017.10.014

Wang L, Zhang W, Chen F (2019) Bayesian approach for predicting soil-water characteristic curve from particle-size distribution data. Energies 14:2992. https://doi.org/10.3390/en12152992

Wang Y, Aladejare AE (2016) Bayesian characterization of correlation between uniaxial compressive strength and Young's modulus of rock. Int J Rock Mech Min Sci 85:10–19. https://doi.org/10.1016/j.ijrmms.2016.02.010

Wang Y, Cao Z (2013) Probabilistic characterization of young's modulus of soil using equivalent samples. Eng Geol 159:106–118. https://doi.org/10.1016/j.enggeo.2013.03.017

Wong TT (2015) Performance evaluation of classification algorithms by k-fold and leave-one-out cross validation. Pattern Recognit 48:2839–2846. https://doi.org/10.1016/j.patcog.2015.03.009

Xia Y, Liu C, Li YY, Liu N (2017a) A boosted decision tree approach using Bayesian hyper-parameter optimization for credit scoring. Expert Syst Appl 78:225–241. https://doi.org/10.1016/j.eswa.2017.02.017

Xia Y, Liu C, Li YY, Liu N (2017b) A boosted decision tree approach using Bayesian hyper-parameter optimization for credit scoring. Expert Syst Appl 78:225–241. https://doi.org/10.1016/J.ESWA.2017.02.017

Yousefpour N, Medina-Cetina Z, Jahedkar K et al (2011) Determination of unknown foundation of bridges for scour evaluation using artificial neural networks. Am Soc Civ Eng pp 1514–1523

Zhang L, Wu X, Liu W, Skibniewski MJ (2019) Optimal strategy to mitigate tunnel-induced settlement in soft soils: simulation approach. J Perform Constr Facil 33:04019058. https://doi.org/10.1061/(asce)cf.1943-5509.0001322

Zhang W, Goh ATC, Zhang Y et al (2015) Assessment of soil liquefaction based on capacity energy concept and multivariate adaptive regression splines. Eng Geol 188:29–37. https://doi.org/10.1016/j.enggeo.2015.01.009

Zhang WG, Goh ATC (2013) Multivariate adaptive regression splines for analysis of geotechnical engineering systems. Comput Geotech 48:82–95. https://doi.org/10.1016/j.compgeo.2012.09.016

Zhang Y, Dai M, Ju Z (2016) Preliminary discussion regarding svm kernel function selection in the twofold rock slope prediction model. J Comput Civ Eng 30:04015031. https://doi.org/10.1061/(asce)cp.1943-5487.0000499

Zhou J, Li E, Wei H et al (2019) Random forests and cubist algorithms for predicting shear strengths of rockfill materials. Appl Sci 9:1621. https://doi.org/10.3390/app9081621

Zhou J, Li X, Mitri HS (2016) Classification of rockburst in underground projects: comparison of ten supervised learning methods. J Comput Civ Eng 30:04016003. https://doi.org/10.1061/(asce)cp.1943-5487.0000553

Chapter 8
Prediction for TBM Penetration Rate Using Four Hyperparameter Optimization Methods and RF Model

8.1 Introduction

With the continuous development and progress of technology, the emergence of tunnel boring machine (TBM) provides a safer, more effective and cheaper construction method for tunnel digging. Because of these advantages, TBM technology has been widely used all over the world. In order to reduce project costs and construction risks, it is very necessary to make an accurate assessment of TBM tunneling efficiency (Gong and Zhao 2009; Mahdevari et al. 2014; Armaghani et al. 2017; Adoko et al. 2017; Liu et al. 2018; Zhang and Gao 2018; Tan et al. 2020) among which, the rate of penetration (ROP) is a key assessment factor (Alvarez Grima et al. 2000). However, for most engineering projects, complex geological conditions make ROP prediction full of uncertainty and randomness. It is generally believed that during tunnel construction, the geological environment, TBM parameters and tunnel geometric conditions will have important impact on ROP (Blindheim et al. 1979; Alvarez Grima et al. 2000; Barton 2000; Sapigni et al. 2002; Yagiz and Karahan 2011; Xu et al. 2019; Zhang et al. 2020). How to determine the impact of each potential parameter on ROP and analyze the relative importance of each parameter have great practical engineering significance. Early prediction models are mostly based on experience, theory or a combination of the two (Bruland 1999; Sapigni et al. 2002; Kahraman et al. 2003; Hassanpour et al. 2009; Yagiz and Karahan 2011; Li et al. 2016, 2020; Xiong et al. 2017; Koopialipoor et al. 2019; Zhang et al. 2020; Yan and Jiang 2020; Hou et al. 2020; Zhu et al. 2020; Yan et al. 2021). Based on the monitoring data of three tunnels, Sapigni et al. (Yagiz 2008) found that ROP is closely related to rock mass quality classification (RMR), and the two can be approximated by quadratic multiple regression equation. Kahraman et al. (Breiman 2001) established a ROP regression model based on field measured data, and found through statistical analysis that ROP is closely related to rock properties. Hassanpour et al. (Breiman 2001) established empirical formulas for ROP and different geological parameters based on Nowsood Tunnel data. The results showed that ROP was closely related to rock tangibility parameters, especially Field penetration index (FPI). In addition,

W. Zhang et al., *Application of Soft Computing, Machine Learning, Deep Learning and Optimizations in Geoengineering and Geoscience*,
https://doi.org/10.1007/978-981-16-6835-7_8

this study shows the convenience of using common parameters such as UCS and RQD to establish the prediction formula. However, empirical formulas are generally established on the basis of local data and are only suitable for a few projects, so they have limited applicability (Breiman 2001). Numerical analysis is also widely used for such problems. (Gong and Zhao 2009) used UDEC software to perform numerical simulation of TBM tool breakage process in two tunnels in Singapore, and found that the tunneling speed of TBM would increase with the increase of rock brittleness index. Statistical analysis of real data also shows a similar view that rock brittleness, as an important mechanical parameter, plays an important role in TBM excavation process. (Li et al. 2019) used Discrete element method (DEM) to study the relationship between ROP and wedge Angle, constraint stress and other parameters between rock bits. The results showed that large wedge Angle would increase the lateral failure of rock, thus reducing the driving speed. In addition, high in-situ stress will also increase the ductility of rock, which is not conducive to excavation. Numerical modeling is very dependent on constitutive model, material parameters and boundary conditions, but due to the complexity and uncertainty of tunnel engineering, these factors are often difficult to obtain accurately. Therefore, numerical analysis can only reflect the general trend, but cannot achieve accurate prediction. In recent years, due to the development of Artificial Intelligence (AI) technology, many data-driven models have been successfully used for ROP prediction. Compared with traditional methods, AI models can capture the complex nonlinear relationship between variables to a considerable extent and make more accurate predictions. (Xu and Niu 2018; Chen et al. 2019; Xu et al. 2019) proposed five different machine learning methods to predict ROP based on the field and laboratory experimental data of a tunnel in Malaysia. The main input parameters used are Uniaxial compressive strength (UCS), Brazilian tensile Strength (Brazilian tensile Strength), BTS, Rock quality designation (RQD), Weathering zone (WZ), Thrust Force (TF) and Revolution per minute (RPM). By comparing the prediction results with the real values, it is found that the K-nearestneighbor (KNN) algorithm can achieve the best prediction accuracy. In addition, the parameter sensitivity analysis found that UCS and RPM had the most important effect on ROP. Taking a tunnel in Malaysia as an example, (Jahed Armaghani et al. 2017) established a ROP prediction model based on Artifical neural network (ANN), and used two optimization algorithms (POS and ICA) to perform hyperparameter tuning. The results show that the prediction accuracy of the optimized model is much higher than that of the ordinary model, which proves the necessity of the optimized algorithm. At present, there are many studies on applying AI models to ROP prediction, but there are still few studies on applying optimization algorithms to related models for hyperparameter tuning to solve the balance between computational accuracy and computational efficiency (Eberhart and Kennedy 1995; Fabian Pedregosa et al. 2011; Mahdevari et al. 2014; Zafar et al. 2017; Mahmoodzadeh et al. 2020; Zhang et al. 2020). In this paper, according to the published data set (Reshef et al. 2011), a random forest (RF) model was proposed to predict ROP, and four common optimization algorithms (PSO, DE, GA and BO) were used to optimize the model hyperparameters. Finally, the optimal model was used to output the relative importance of parameters.

8.2 Engineering Background and Data Introduction

The data set used in this paper is from literature (Reshef et al. 2011). The data set consists of 153 data samples from 151 different locations of the tunnel, including 7 characteristic components including rock geological parameters, tunnel geometry conditions and tunneling speed of road header. As the RF model itself is suitable for scenarios with low data dimension and high accuracy requirements, the data set used in this paper has certain representativeness and guiding significance for similar engineering projects.

8.2.1 Project Description and Geological Survey

New York City built phase 2 of the Queens Water Tunnel to improve the distribution and transportation of fresh water in the city. The tunnel is 7.5 km long, 7.06 m in diameter and about 200 m deep. The construction process was completed by a powerful tunnel boring machine (TBM). The geological structure of the construction area along the tunnel is very complex, which is composed of different metamorphic igneous rocks and has local weak structural planes such as shear zone, joint and fault. The rock composition mainly consists of rhyolite acite, granite gneiss, amphibolite, igneous gneiss and schist, and the respective proportions are shown in (Barton 2000).

The data set was sampled from 151 different locations in the tunnel, including 153 samples. The sampled cores were ROP of the TBM at each sampling point was also analyzed based on the measured data. Therefore, the whole data set includes the strength and brittleness and toughness characteristics of the rock. The strength characteristics are reflected by BTS and UCS, while the brittleness and toughness are calibrated by the impact penetration experiment. For the impact penetration experiment, the ratio of the maximum load the core can bear to the corresponding maximum displacement (Peak slope index, PSI) is often used to reflect the brittleness and toughness of the rock sample.

In addition, in order to reflect the influence of the type and density of rock discontinuity on the tunneling speed of TBM, the included Angle between the tunnel axis and the soft surface was also calculated and recorded in the data set. According to the classification system developed by NTNU, rock-fracture class (RC) can be divided into 7 categories according to their integrity, which are numbered from 1 to 7. The larger the number is, the more weak structural planes of Rock mass are, and the more fractured the Rock mass is. Meanwhile, the data set also records the Average distance between planes of Weakness (DPW) corresponding to the classification results.

8.2.2 Data Description

Figure 8.1 shows the distribution characteristics of the above variables and the linear correlation between the variables. The diagonal shows the distribution characteristics of each variable; The left diagonal indicates the linear relationship among variables, and the fitting curve represents the degree of correlation. The higher the degree of inclination, the stronger the correlation. On the right of the diagonal, the statistical characteristics of each variable are shown, including the information of mean value, standard deviation, maximum value, minimum value and median. From the linear relationship, we can see that there is a certain correlation between the variables. For the prediction target value ROP, it can be seen from the correlation that it has a strong positive correlation with the input variables PSI and RC, and a strong negative correlation with DPW, and a weak correlation with other variables.

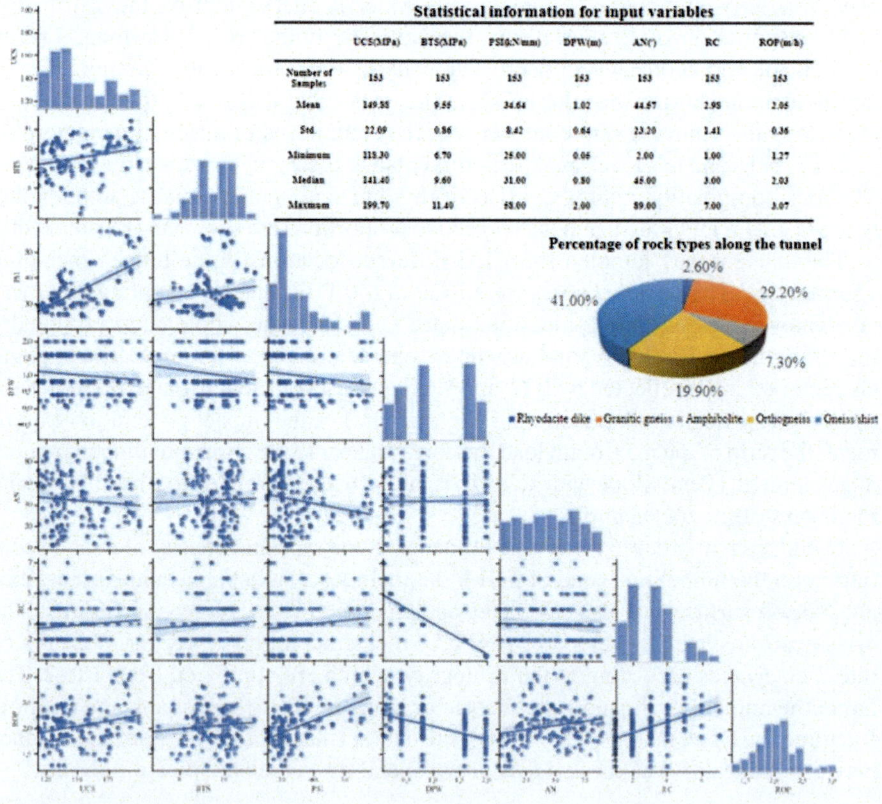

Fig. 8.1 Description of correlation and statistical features of variables and percentage of tunnel rock classification

In fact, the driving speed of TBM is a complex dynamic nonlinear system, which is related to many factors. Linear correlation cannot fully reflect the degree of influence of variables on ROP, so it is necessary to introduce nonlinear evaluation indexes to further illustrate the correlation. The concept of Maximal information coefficient (MIC) was first established by (Reshef et al. 2011), which can be used to comprehensively evaluate the linear or nonlinear strength between two variables, and belongs to the maximum information based non-parametric exploration class of statistics. Its basic principle is to discretization the relationship between two variables into a two-dimensional space and represent it with a scatter diagram. Then the current two-dimensional space is divided into certain interval numbers along the X and Y directions respectively, and then the situation of the scatter falling into each square is checked, that is, the calculation of joint probability. After the maximum value of mutual information is obtained, it is normalized, and then the maximum value of mutual information at different scales is compared as MIC value to reflect the relationship between variables.

As can be seen from Fig. 8.2, the driving speed ROP has a certain nonlinear correlation with UCS, PSI, DPW and RC, but has a general correlation with BTS.

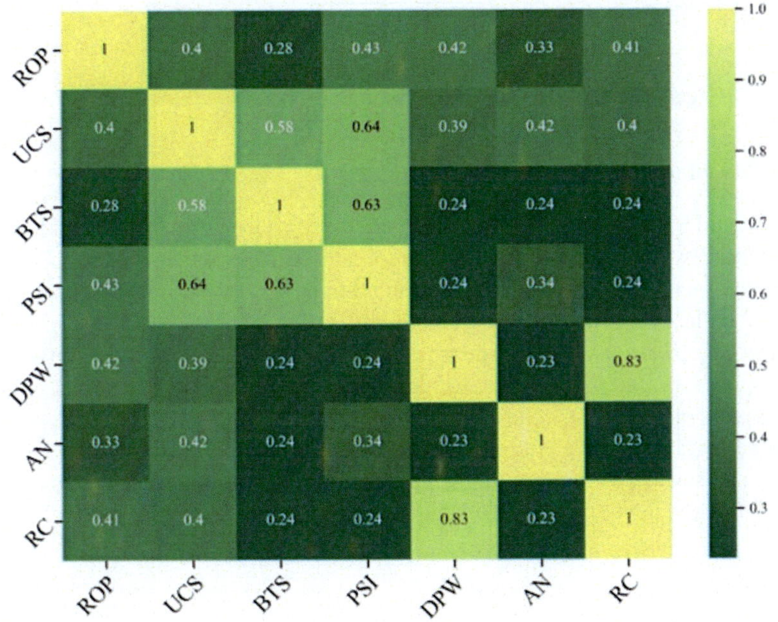

Fig. 8.2 The nonlinear correlation between variables expressed by the maximal information coefficient (MIC)

8.3 Prediction and Sensitivity Analysis of Driving Speed

Before the data set is broken up, the data is normalized to the interval of [-1,1], which will help the convergence of the model and improve the calculation efficiency. The specific operations are as follows:

$$X_{new} = 2 * \frac{X_i - X_{mean}}{X_{\max} - X_{\min}} - 1 \tag{8.1}$$

where, is the normalized data, is the data to be normalized, and is the maximum and minimum values in the data set. After normalization, the original data set was randomly divided into training set and test set at the ratio of 80% and 20%. The former was used for 5-recross validation method and optimization algorithm, and the related prediction model was established. The latter is used as a new data set to test the performance of the optimized model. After the division, the statistical characteristics of ROP are shown in Table 8.1. It can be seen that the statistical indicators of the two data sets have little difference, which can avoid the phenomenon of overfitting or underfitting to some extent.

8.3.1 *Prediction Model*

(1) Model establishment

UCS, BTS, PSI, DPW, AN and RC were taken as input variables, and the driving speed ROP was taken as output target to establish the related model. When PSO optimization is adopted, the particle population is 30, the maximum iteration number is 100, the learning factor c1 and c2 are both 0.5, the inertia weight w is 1, the tree number es is [0,50], and the tree depth dp is [0,50]. When GA optimization is adopted, the population number is set to 30, the maximum number of iterations is 200, the error precision is 1×10^{-7}, the tree number es is [0,50], the tree depth dp is [0,50], and the other parameters are default. When DE optimization is adopted, the population number is set to 30, the maximum iteration number is 200, the tree number es is [0,50], the tree depth dp is [0,50], and the other parameters are default values. When

Table 8.1 Statistical characteristics of ROP in training and test set

	Training	Testing
Num	122	31
Mean	2.06	1.98
Std	0.36	0.33
Max	3.07	2.95
Min	1.27	1.51

Table 8.2 The optimal parameters and time cost of each random forest model after optimization

Model	Tree number (es)	deep(dp)	Time (s)
PSO-RF	49	38	144.82
GA-RF	35	40	281.67
DE-RF	37	16	430.66
BO-RF	17	37	66.70

BO optimization is adopted, the initial node is 30, the maximum iteration number is 200, the tree number es is [0,50], the tree depth dp is [0,50], and the other parameters are default (Table 8.2).

(2) Model performance comparison

The prediction results of different models for the training set and the test set are shown in Fig. 8.3. The reference line represents the optimal situation, that is, the predicted value is exactly equal to the real value. Therefore, the closer the prediction result is to the reference line, the better the prediction effect is; otherwise, the worse the prediction effect is. On the whole, it can be seen that there is no significant difference between the model performance, indicating that the test results are reasonable. For each model, the comprehensive performance of BO-RF model is the best. As for the correlation coefficient R, it is the maximum in both the training set and the test set, indicating that the prediction results are highly correlated with the real results. While the de-RF model has a relatively poor performance. Although the results are similar in the training set, the correlation coefficient is slightly smaller than the other three models in the test set. The performance of GA-RF and PSO-RF models was basically the same, and the evaluation parameters were almost the same.

If the calculation time is taken into account, it can be seen that the BO-RF model is still the best model, achieving the highest calculation accuracy in the least time. However, the de-RF model has a poor performance, which uses the most time to calculate the lowest accuracy. Although the PSO-GA model takes less time than the GA-RF model, considering that the model performance is basically the same, the PSO-RF model is more recommended than the GA-RF model.

Figure 8.4 shows the performance of the different RF models on the test set, which again proves that the predictive performance of the models is close to the same. It can be seen from the absolute error distribution that the maximum error between the predicted value and the true value of each model is less than 1 m/hour, indicating that each model can control the error well. Compared wit models cannot predict some mutation points very well, partly because there is a certai h other models, the BO-RF model has a smaller error distribution. However, all the n error in the calculation of the true value, and partly because the correlation between input variables and output targets is not very strong.

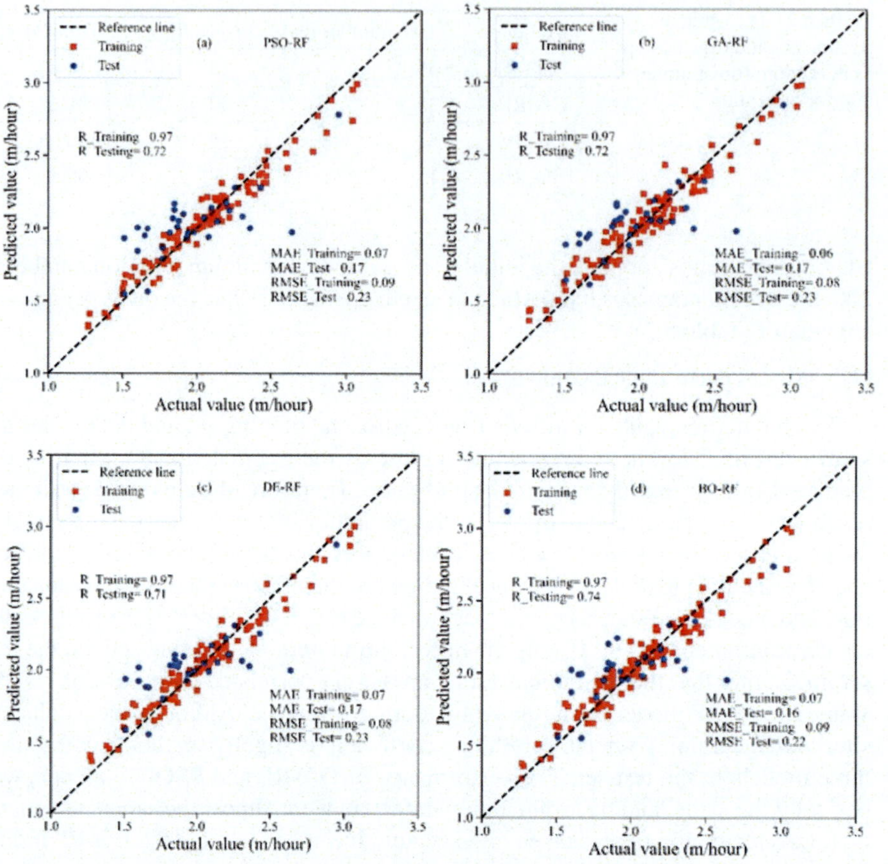

Fig. 8.3 Best prediction results of different RF models on the dataset

8.3.2 Sensitivity Analysis

Because the bo-RF model has the best calculation accuracy and the least time consuming, the sensitivity analysis results are analyzed by the Bo-RF model. Figure 8.5 shows the relative importance of each input variable of the output of the BO-RF model. It can be seen that UCS and BTS play an absolute control role on the model, while RC has almost no influence on the model. The other three variables have some influence on the performance of the model. It is worth noting that this result is more close to the nonlinear analysis result in Sect. 7.3.2, which also indicates that the TBM driving speed is a complex nonlinear system, which is strongly correlated with the independent or coupled effects of various factors. In addition, the analysis results of this paper are basically consistent with the results of similar literature (Zhang et al. 2020), indicating the accuracy of the prediction model and providing relevant reference for practical engineering. Due to the limitation of the

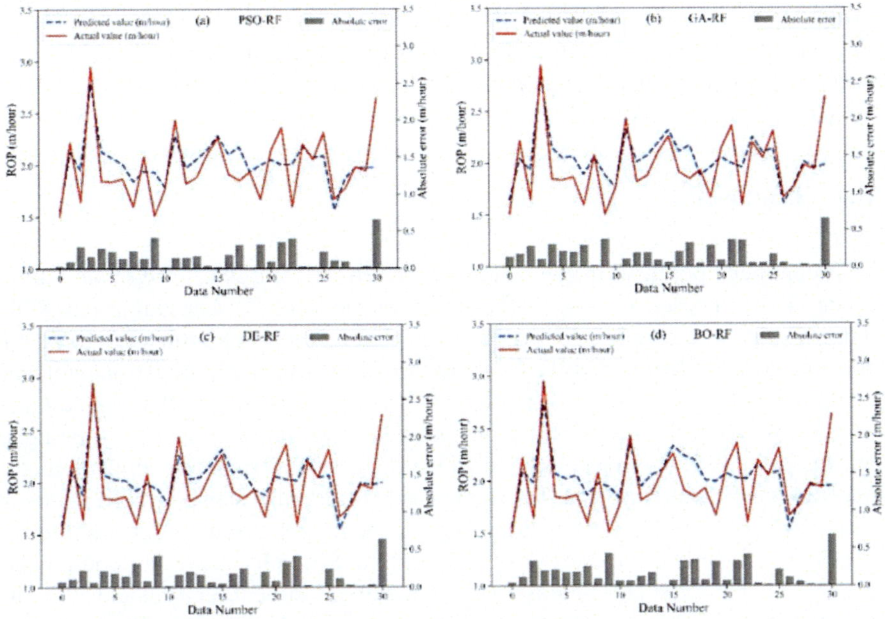

Fig. 8.4 Best prediction results of different RF models on the dataset

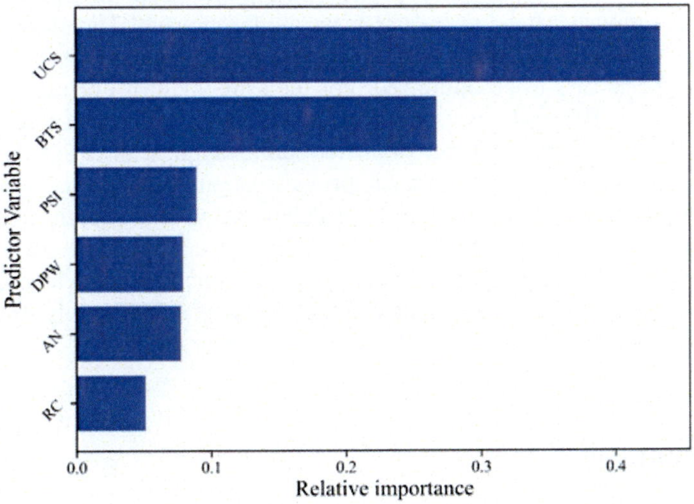

Fig. 8.5 Results of sensitivity analysis

sample size of the data set, it can be considered to collect more data to update and upgrade the model in the later stage, so as to obtain a more comprehensive and accurate parameter relative importance.

8.4 Discussion

Four common optimization algorithms (PSO, GA, DE and PSO) and random forest model (RF) were used to construct the prediction model of TBM driving speed (ROP). The rock sample strength index (UCS, BTS), the impact penetration test index (PSI), the tunnel rock integrity (DPW, RC) and the tunnel and shield direction geometric condition (AN) are taken as the input, and the driving speed (ROP) as the output.

By comparing the performance of different optimization models in training set and test set, it is found that bo-RF model has the best performance in both prediction accuracy and calculation time. The main reason is that Bayesian optimization adopts Gaussian process (GP), which can take into account the previous parameter information and constantly update the prior information. Secondly, Bayesian optimization is very suitable for small-dimensional parameter adjustment requirements, because the number of iterations is small and the convergence speed is fast. However, other optimization algorithms adopt gradient descent method, which is slow in operation and can easily lead to dimension explosion when multiple parameter dimensions are used, thus seriously affecting the parameter tuning process. Therefore, for the RF model in this paper, BO optimization shows certain advantages because there are only two parameters to be optimized. For the same kind of practical engineering problems, when RF model is used for prediction, BO algorithm can be used as the preferred term to carry out hyperparameter optimization, while PSO algorithm can be used as the standby term to compare the calculation results and calculation costs. It should be noted that RF model is applicable to small and medium scale data and high prediction accuracy requirements, and has a certain range of use. In addition, in order to avoid the phenomenon of overfitting, common outlier detection methods, such as methods, are used to eliminate outliers, which can effectively strengthen the stability of the model and improve the prediction accuracy.

8.5 Summary

Aiming at the complex nonlinear dynamic system of tunnel tunneling speed, a random forest prediction model based on four different optimization algorithms is proposed. Will be used to predict the model Queens, New York area water conveyance tunnel drivage rate, although the complex geological conditions, tunnel construction area model input parameters, less samples, but the model prediction and the actual value with high consistency, and the parameters of the model output relative importance and prophase correlation analysis and the results are basically identical documents, shows that reliability and applicability of the model. Through the analysis and comparison of the results, it can be concluded that the BO-RF model has the best performance in capturing the complex nonlinear relationship between the potential influencing factors and TBM tunneling speed. This model has high accuracy and less time consuming, so it is recommended to be applied to the prediction of such problems. The performance of PSO-RF model is similar to that of GA-RF model, but the former takes less time and is more efficient. The de-RF model takes the most time, but the performance of the model does not improve significantly. The prediction of TBM tunneling speed by this method remains to be studied. The sensitivity analysis results show that the influence of each input parameter on the TUNNELING speed of TBM is different, and the strength parameter of rock itself is the absolute influence. As a mature machine learning method, random forest model has been widely used in practical engineering problems because of its low number of hyperparameters to be optimized, high computational efficiency and the ability to avoid overfitting problems. However, the size and quality of the data also have a direct impact on the output results of the model. Therefore, more data samples can be added in the future to improve the performance of the optimized model and conduct a comprehensive and in-depth study on the correlation of input parameters. In addition, the development of predictive models relies heavily on hyperparametric optimization, which may require expertise and historical experience, and the "black box" nature of some intelligent models makes them less interpretable and needs to be applied according to the needs of the problem.

References

Adoko AC, Gokceoglu C, Yagiz S (2017) Bayesian prediction of TBM penetration rate in rock mass. Eng Geol 226:245–256. https://doi.org/10.1016/j.enggeo.2017.06.014

Alvarez Grima M, Bruines PA, Verhoef PNW (2000) Modeling tunnel boring machine performance by neuro-fuzzy methods. Tunn Undergr Sp Technol 15:259–269. https://doi.org/10.1016/S0886-7798(00)00055-9

Armaghani DJ, Mohamad ET, Narayanasamy MS et al (2017) Development of hybrid intelligent models for predicting TBM penetration rate in hard rock condition. Tunn Undergr Sp Technol 63:29–43. https://doi.org/10.1016/j.tust.2016.12.009

Barton N (2000) TBM tunnelling in jointed and fault.pdf. Balkema

Breiman L (2001) Random forests. Mach Learn 45:5–32. https://doi.org/10.1023/A:1010933404324

Bruland A (1999) Hard rock tunnel boring advance rate and cutter wear. Trondheim Nor Inst Technol 3:Project report 1B-98, NTNU Trondheim

Chen W, Hong H, Panahi M et al (2019) Spatial prediction of landslide susceptibility using GIS-based data mining techniques of ANFIS with whale optimization algorithm (WOA) and grey wolf optimizer (GWO). Appl Sci 9:. https://doi.org/10.3390/APP9183755

Drillability predictions in hard rock tunnelling: Blindheim, O T In: Tunnelling \"79, Proceedings of the 2nd international symposium, London, 12–16 March 1979, P284–289. Publ London, IMM, 1979

Eberhart R, Kennedy J (1995) New optimizer using particle swarm theory. In: Proceedings of the international symposium on micro machine and human science. IEEE, pp 39–43

Gong QM, Zhao J (2009) Development of a rock mass characteristics model for TBM penetration rate prediction. Int J Rock Mech Min Sci 46:8–18. https://doi.org/10.1016/j.ijrmms.2008.03.003

Hassanpour J, Rostami J, Khamehchiyan M, Bruland A (2009) Developing new equations for TBM performance prediction in carbonate-argillaceous rocks: a case history of nowsood water conveyance tunnel. Geomech Geoengin 4:287–297. https://doi.org/10.1080/17486020903174303

Hou S, Liu Y, Zhang K (2020) Prediction of TBM tunnelling parameters based on IPSO-BP hybrid model. Yanshilixue Yu Gongcheng Xuebao/Chinese J Rock Mech Eng 39:1648–1657. https://doi.org/10.13722/J.CNKI.JRME.2019.1084

Jahed Armaghani D, Shoib RSNSBR, Faizi K, Rashid ASA (2017) Developing a hybrid PSO–ANN model for estimating the ultimate bearing capacity of rock-socketed piles. Neural Comput Appl 28:391–405. https://doi.org/10.1007/s00521-015-2072-z

Kahraman S, Bilgin N, Feridunoglu C (2003) Dominant rock properties affecting the penetration rate of percussive drills. Int J Rock Mech Min Sci 40:711–723. https://doi.org/10.1016/S1365-1609(03)00063-7

Koopialipoor M, Tootoonchi H, Jahed Armaghani D et al (2019) Application of deep neural networks in predicting the penetration rate of tunnel boring machines. Bull Eng Geol Environ 78:6347–6360. https://doi.org/10.1007/S10064-019-01538-7

Li SH, Wu LZ, Luo XH (2020) A novel method for locating the critical slip surface of a soil slope. Eng Appl Artif Intell 94. https://doi.org/10.1016/J.ENGAPPAI.2020.103733

Li X, Liu X, Li CZ et al (2019) Foundation pit displacement monitoring and prediction using least squares support vector machines based on multi-point measurement. Struct Heal Monit 18:715–724. https://doi.org/10.1177/1475921718767935

Li XF, Li HB, Liu YQ et al (2016) Numerical simulation of rock fragmentation mechanisms subject to wedge penetration for TBMs. Tunn Undergr Sp Technol 53:96–108. https://doi.org/10.1016/J.TUST.2015.12.010

Liu X, Zhou S, Xu M, et al (2018) Analysis of rock fragmentation process with double circle cutters in mixed ground. Yingyong Jichu yu Gongcheng Kexue Xuebao/Journal Basic Sci Eng 26:357–370. https://doi.org/10.16058/j.issn.1005-0930.2018.02.013

Mahdevari S, Shahriar K, Yagiz S, Akbarpour Shirazi M (2014) A support vector regression model for predicting tunnel boring machine penetration rates. Int J Rock Mech Min Sci 72:214–229. https://doi.org/10.1016/j.ijrmms.2014.09.012

Mahmoodzadeh A, Mohammadi M, Daraei A et al (2020) Forecasting maximum surface settlement caused by urban tunneling. Autom Constr 120. https://doi.org/10.1016/j.autcon.2020.103375

Pedregosa F, Varoquaux G, Gramfort A et al (2011) Scikit-learn: machine learning in python. J Mach Learn Res 12:2825–2830

Reshef DN, Reshef YA, Finucane HK, et al (2011) Detecting novel associations in large data sets. Science (80-) 334:1518–1524. https://doi.org/10.1126/science.1205438

Sapigni M, Berti M, Bethaz E et al (2002) TBM performance estimation using rock mass classifications. Int J Rock Mech Min Sci 39:771–788. https://doi.org/10.1016/S1365-1609(02)00069-2

Tan H, Ji HG, Zeng ZY, Liu ZQ (2020) Optimal drilling pressure of cone-tipped cutters based on characteristic size of hard and brittle rocks. Yantu Gongcheng Xuebao/Chinese J Geotech Eng 42:782–789. https://doi.org/10.11779/CJGE202004023

Xiong F, Hu Z, Ren X, Zhang P (2017) Matlab-based bp neural network applied to the prediction of tbm advance rate. Mod Tunn Technol 54:101–107. https://doi.org/10.13807/J.CNKI.MTT.2017.05.014

Xu H, Zhou J, Asteris PG, et al (2019) Supervised machine learning techniques to the prediction of tunnel boring machine penetration rate. Appl Sci 9. https://doi.org/10.3390/app9183715

Xu S, Niu R (2018) Displacement prediction of Baijiabao landslide based on empirical mode decomposition and long short-term memory neural network in Three Gorges area, China. Comput Geosci 111:87–96. https://doi.org/10.1016/j.cageo.2017.10.013

Yagiz S (2008) Utilizing rock mass properties for predicting TBM performance in hard rock condition. Tunn Undergr Sp Technol 23:326–339. https://doi.org/10.1016/j.tust.2007.04.011

Yagiz S, Karahan H (2011) Prediction of hard rock TBM penetration rate using particle swarm optimization. Int J Rock Mech Min Sci 48:427–433. https://doi.org/10.1016/j.ijrmms.2011.02.013

Yan CB, Wang HJ, Yang JH et al (2021) Predicting TBM penetration rate with the coupled model of partial least squares regression and deep neural network. Yantu Lixue/Rock Soil Mech 42:519–528. https://doi.org/10.16285/j.rsm.2020.0164

Yan C, Jiang X (2020) Prediction model of tbm net advance rate based on parameters of rock mass and tunnelling. Mod Tunn Technol 57:26–33. https://doi.org/10.13807/J.CNKI.MTT.2020.02.004

Zafar A, Shah S, Khalid R, et al (2017) A meta-heuristic home energy management system. In: Proceedings—31st ieee international conference on advanced information networking and applications workshops, WAINA 2017. pp 244–250

Zhang W, Tang L, Li H et al (2020) Probabilistic stability analysis of Bazimen landslide with monitored rainfall data and water level fluctuations in Three Gorges Reservoir, China. Front Struct Civ Eng 14:1247–1261. https://doi.org/10.1007/s11709-020-0655-y

Zhang Z, Gao Q (2018) Theory and Application of Full Face Rock Tunnel Boring Machine Cutter-head Bending Model. Yingyong Jichu yu Gongcheng Kexue Xuebao/Journal Basic Sci Eng 26:1121–1129. https://doi.org/10.16058/J.ISSN.1005-0930.2018.05.018

Zhu M, Zhu H, Wang X, Cheng P (2020) Study on CART-based ensemble learning algorithms for predicting TBM tunneling parameters and classing surrounding rockmasses. Yanshilixue Yu Gongcheng Xuebao/Chinese J Rock Mech Eng 39:1860–1871. https://doi.org/10.13722/J.CNKI.JRME.2019.0924

Chapter 9
What We Have Learnt from the Applications

It is well known that the use of SC and more advanced AI techniques including ML, DL and OA has greatly improved the accuracy and efficiency of prediction for the responses of geotechnical engineering structures. From Chap. 5, it can be known that compared with tunneling and braced excavations, AI algorithms are rarely used in the field of piles and pipelines. Thus, it is suggested that more algorithms need to popularize their use in these two areas in the future. Also, based on the literatures collected, it is obvious that single algorithm-based models are widely used in the geotechnical engineering, such as DT, SVM, RF, ANN, BPNN. The popularity of ensemble or hybrid models and optimization algorithms deserve to be further improved. Figure 9.1 briefly provides the ML recent R&D as well as the main challenges and limitations.

Besides, the traditional physical-based models are built based on the known underlying relationship between the geotechnical structures' responses and the influential factors. However, the AI-based models just establish a direct mapping between inputs and outputs and cannot provide any explicit expressions. Therefore, coupling the use of physics-based and AI data-driven models deserves more attention in the future research. In addition, Table 9.1 lists the merits and disadvantages of each method mostly mentioned above.

Natural hazards or disasters pose a threat to people's lives, properties, and the environments. Many researchers have attempted to mimic and better understand these disasters, using empirical equations, laboratory experiments, numerical simulations. They are of certain merits, on the other hand, they suffer from the constraints or limitations for further application. Recently, AI methods have gained increasing popularity, as they can mimic the complex mathematical expressions of physical processes of natural disasters, providing better performance and cost-effective solutions. However, AI methods are not perfect either, as they rely heavily on the quality and quantity of collected data. It can be guessed that if a ML model performs very well on a given dataset, it still has the possibility to perform poorly on other datasets. Besides, given a specific geo-scientific problem, there are many types of ML methods available. Therefore, how to choose a suitable method in emergent condition and how

W. Zhang et al., *Application of Soft Computing, Machine Learning, Deep Learning and Optimizations in Geoengineering and Geoscience*,
https://doi.org/10.1007/978-981-16-6835-7_9

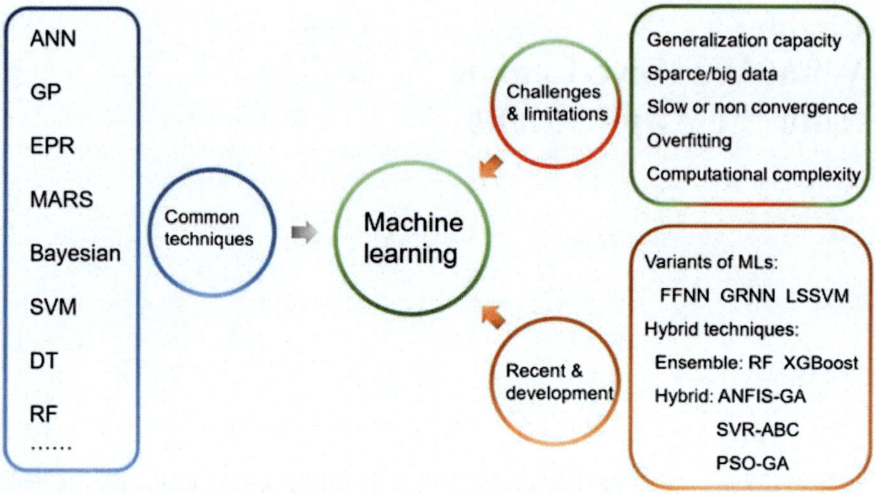

Fig. 9.1 Recent ML R&D and the main challenges and limitations

to optimize the hyper-parameters correctly requires certain professional skills. This may limit the applicability of ML methods to some extent. In addition, the black-box nature of ML methods makes it impossible to explain the physical processes of the output, and this may be unacceptable for decision-makers, as ML methods are not transparent and interpretable.

Considerable monitoring work using SAR, INSAR and LiDAR has been performed against geohazards especially for landslides which currently has attracted increasing attention to the disaster prevention and reduction, disaster resistance and relief, but more should be done to investigate how to detect potential landslides from these big data in a consistent, reliable, and smart manner. ML and DL have been widely implemented in the field of computer science and remote sensing, where statistical techniques are employed to learn specific and complex tasks from given data. Recent studies report that ML can identify signals associated with geohazards from large data sets. However, predicting when landslides will occur remains a great challenge, considering that there have been only a limited number of successful case studies reported. Figure 9.2 summarizes the main constrained further applications of DL in geoengineering and geosciences.

Table 9.1 Pros and cons of ML, DL and OA mentioned in this book

Method	Sub-type	Algorithm	Pros	Cons
ML	**Supervised learning**	**ANN**	• works well even with insufficient knowledge • distributed memory • parallel processing	• hardware dependence • lack of interpretation
		ELM	• fast to train • easy to use • greater generalization performance with appropriate activation function	• cannot be used in complex works • its randomness often causes additional uncertainty • inappropriate activation function would cause generalization degradation
		DT	• good interpretability • less effort for data preparation required during pre-processing • handles both classification and regression problems	• high probability of overfitting
		Linear Regression	• works well with small datasets • outputs probability directly • simplicity and transparency	• cannot work for non-linear problems
		GP	• the length of solution can be extended up to hardware limits • users do not need to have maximum knowledge of the problem and their solutions	• small variation often cause disastrous effect on fitness of the solution generated • the program will be slow and errors will be generated if any of the used high level language needs to be compile

(continued)

Table 9.1 (continued)

Method	Sub-type	Algorithm	Pros	Cons
		NB	• works well with small datasets • handles both continuous and discrete data • fast computing • not sensitive to irrelevant features	• predictor features are assumed to be independent • the assumption on prior probability is required
		SVM	• good performance when there is a clear margin of separation between classes • effective in high dimensional spaces • high memory efficiency • handles both linear and non-linear problems	• not suitable for large datasets • bad performance when there are more noise points • does not directly provide probability estimates
		KNN	• no training period required • adding new data will not impact the accuracy of the algorithm • easy to implement	• does not work well with large datasets • does not work well with high dimensions • sensitive to noisy data points
		XGboost	• high flexibility • parallel processing • prevents overfitting • good interpretability • high accuracy	• does not perform well on sparse and unstructured data • sensitive to outliers • is hardly scalable
	Unsupervised learning	**k-means**	• simple to implement • scales to large datasets • guarantees convergence • easily adapts to new examples	• k needs to be chosen manually • clustering outliers • dependent on initial values • has trouble clustering data where clusters are of varying sizes and density • slow convergence rate on large datasets

(continued)

Table 9.1 (continued)

Method	Sub-type	Algorithm	Pros	Cons
		PCA	• removes correlated features • reduces overfitting • improves visualization	• information loss • interpretability is deteriorated as original variables are not remained at all
DL	/	**AE**	• can learn non-linear transformations • high computational efficiency • pre-trained layers from another model can be used to apply transfer learning to enhance the encoder/decoder	• needs a considerable amount of training data • misunderstanding important variables • too specialized
		DBN	• efficient usage of hidden layers • robustness in classification	• hardware dependence
		CNN	• automatically detects the important features • high accuracy in image recognition problems • performs well in high-dimensional spaces	• requires a large amount of training data • lack of interpretation
		RNN	• Remembers each information throughout the time • can process inputs of any length	• high computation cost • prone to problems such as gradient exploding and gradient vanishing
		LSTM	• capture the sequential information • can process inputs of any length • not vulnerable to vanishing gradient problems	• high probability of overfitting • high computation cost • sensitive to different random weight initialization
OA	**Precise algorithm**	**SGD**	• converges faster for large datasets • avoids local minims	• not suitable for parallel processing • less accurate

(continued)

Table 9.1 (continued)

Method	Sub-type	Algorithm	Pros	Cons
		CG	• stable performance • small amount of memory required • faster convergence	• not suitable for deep learning
		GA	• avoids being trapped in local optimal solutions • supports multi-objective optimization • works well on both discrete and continuous problems	• computationally expensive • difficult to represent and operate
	Evolutionary Algorithm	**Bayesian Optimization**	• needs only a small number of evaluation times • time saving	• the assumption on prior probability is required
		PSO	• easy to implement • able to run parallel computation • converge fast	• difficult to define initial design parameters • not suitable for solving scattering problems • can be trapped into a local optima in complex problems
		DE	• simple to operate • strong global convergence and robustness	• not suitable for solving scattering problems
		ABC	• simple to implement • strong flexibility and robustness • broad applicability	• lack of use of secondary information • requires new fitness tests on new algorithm parameters • slow when in sequential processing

(continued)

Table 9.1 (continued)

Method	Sub-type	Algorithm	Pros	Cons
		ACO	• search among a population in parallel • guarantees convergence • rapid discovery of good solutions	• uncertain time to convergence • difficult theoretical analysis
		CS	• less complex structure	• cannot converge to the global optima when dealing with complex problems
		FA	• auto-segmentation to avoid local minims • deals with multi-modal functions more efficiently	• its low exploration capability sometimes makes it impossible to achieve the global optimal solution
		GWO	• easy to implement • less storage and computational requirements • ability in avoiding local minim • good stability and robustness	• low solving accuracy • bad local searching ability • slow convergence rate
		WOA	• good stability and robustness	• low solving accuracy • easy to fall into local optimum • slow convergence rate
		SFLA	• powerful in solving complex combinatorial optimization problems • robust in determining the global solution	• slow convergence rate
		CSO	• good at finding global best solution • less time complexity • deal with multi-objective problems	• slow running speed

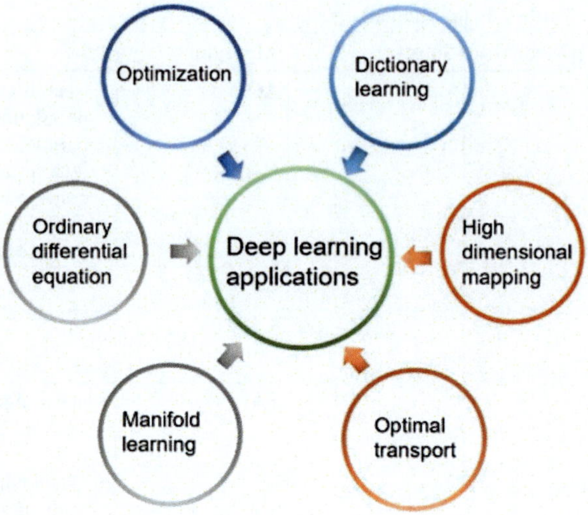

Fig. 9.2 Constraint for further applications of DL in geoengineering and geosciences

Chapter 10
Work Ongoing and Future Recommendations

As far as the authors understand, the following work is going on while some future recommendations might be proposed:

(1) Machine learning models are built on data, and by design they do not incorporate any physical law (such as mass and energy balance) and do not extrapolate well beyond the range of the training data. Historically, physical modelling and machine learning have generally been treated as two different fields or avenues with very different scientific paradigms (theory-driven versus data-driven, as depicted in Fig. 10.1). Yet, in fact these two approaches are complementary, with physical approaches in principle being directly interpretable and offering the potential of extrapolation beyond observed conditions, whereas data-driven approaches are highly flexible in adapting to data and are amenable to finding unexpected patterns. The foremost challenges facing geoengineers and geoscientists is how to extract as much useful information as possible and how to gain new insights from both machine learning and simulations and the interplay between the two. Different techniques such as regularization and embedding the physical constraints into the model architecture have been successfully implemented to address this issue. However, more research and investment in building domain-aware models and using techniques such as reinforcement learning is a high priority (Fig. 10.1).

(2) Based on the machine/deep learning principle of 'Garbage in, garbage out', the selection of training observations directly influences the prediction success or failure. Some deficiencies exist in prediction of geoengineering or geoscience with ML and DL. Firstly, a common problem is the lack of training samples. Training data is the main component of supervised machine learning techniques and is increasingly becoming the main bottleneck to advance applications of ML techniques in geoengineering and geoscience. A small amount of training samples cannot feed enough features into a ML or DL model, which leads to less satisfactory predictive accuracy. Another problem is observation imbalance, which means there are significant differences among the number of labels in

W. Zhang et al., *Application of Soft Computing, Machine Learning, Deep Learning and Optimizations in Geoengineering and Geoscience*,
https://doi.org/10.1007/978-981-16-6835-7_10

Fig. 10.1 Comparison of two alternative avenues: physics-based versus data-driven

the training set. For this regard, while development and sharing of benchmark training datasets is essential, research and investment in techniques that require less training data (such as semi-supervised learning and active learning) is required. In addition, the use of synthetic and simulation datasets to build ML models that mimic the physics of geoengineering and geoscience systems was also recommended. Furthermore, integrating process-based models with ML models will help alleviate data limitations in view that the integration of physical knowledge, process-based models and ML helps to infer unknown quantities.

(3) For DL-based algorithm such as the kernel-based CNN, assessment of landslide or debris flow based on the image identification is generally made up of four steps, which include image preprocessing, segmentation of region of interest, extraction of features and label classification. Since the experiments are done based on different standards of gray-scale INSAR images with different sizes, the INSAR images are corrupted by the so-called salt and pepper noise, speckle noise and random noise produced during imaging preprocessing. In addition, even the high-quality INSAR images are taken, due to the constraint of inherent tempo decorrelation, spatial decorrelation, atmospheric delay, etc., the imaging techniques as well as the standardization involved should be further improved.

(4) For the geoengineering problems, generally the predictions of bearing capacity or the deformation are involved, which are regarded to be of numeric estimation. While for geo science problems, the assessment of susceptibility or the vulnerability of geohazards, or soil/rock type classification are of concern, which are of the categorical nature. However, considering the spatial variability and the anisotropic characteristics of the geomaterials, considerable uncertainty exists in quantifying the stiffness and strength. Therefore, the outputs either via the machine learning or deep learning are not and should not be deterministic. In this regard, the authors hold the view that the combined use of the data-driven ML or DL, the knowledge-based /physics-based approach is essential, a possible working mechanism is provided in Fig. 10.2, as below. Subsequently,

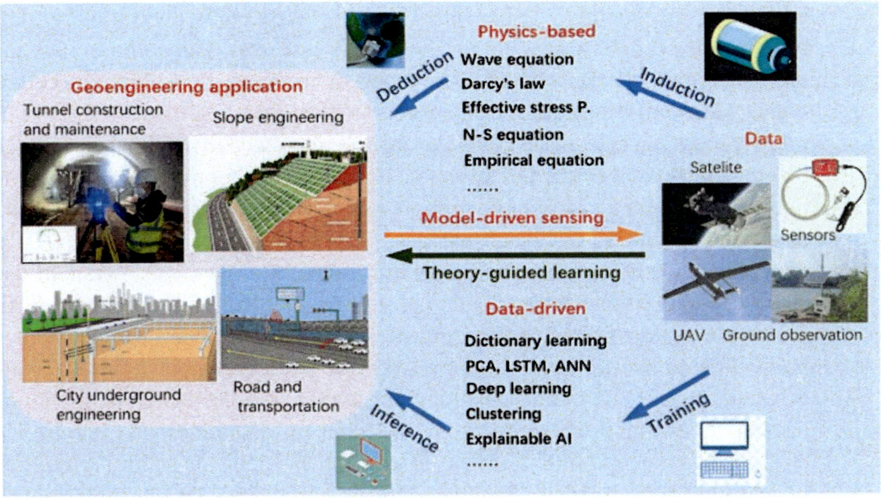

(a)

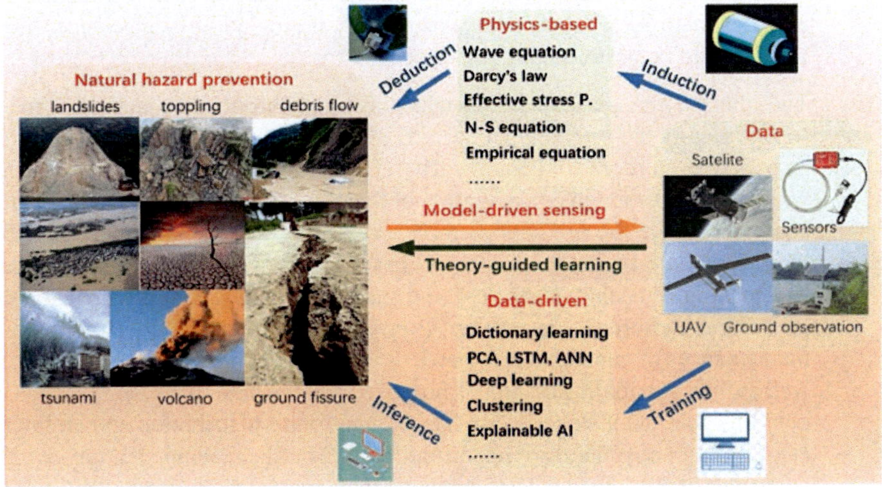

(b)

Fig. 10.2 Possible working mechanism for combined use of the data-driven ML or DL and the knowledge-based /physics-based approach: **a** geoengineering, and **b** geoscience

coupling the data-driven ML or DL models with the physical models to develop the so-called disruptive technology is the cutting-edge technique.

(5) Development of benchmark dataset. Currently there are quite a lot of papers relevant with the machine learning algorithms or soft computing techniques published. Each claims that the proposed methods work better than other approaches. However, this conclusion is arrived based on the different datasets

adopted. Consequently, the benchmark dataset, especially dataset of big data, used for model calibration and verification is essential. The authors think that the benchmark dataset should satisfy one or more of the following requirements: (1) monitoring data over a few years, (2) extensive site investigations, (3) multi-variate big data, (4) available to anyone interested in machine learning. In this regard, several concerted efforts have started in recent years to catalog and publish benchmark datasets to support new model development and running data science challenges, such as "Kaggle Competitions" (https://www.kaggle.com/), "Open Images Challenge"(https://www.ai.google/), "Tianchi data science competition"(https://tianchi.aliyun.com/), "Lotus contests for TBM predictive performances" organized in China, "Data science competitions to build a better world" organized by DRIVENDATA company (https://www.drivendata.org), etc. While development and sharing of benchmark training datasets is essential, research and investment in techniques that require less training data (such as semi-supervised learning and active learning) is required. In addition, data science challenges are essential, however, a major barrier to realizing the benefits of data science and ML lies in our undergraduate and graduate curricula. Little in geotechnical engineering standard curriculum prepares them for a future with substantially more data-driven science. More work should be performed in this aspect.

(6) The climate change and the ever-increasing anthropogenic activities make forecasting of the possible geohazards and the estimation of geoengineering responses even more challenging. Therefore, machine/deep learning methods with modularization in consideration of the climate change and anthropogenic activities enable the predictions more extensible, maintainable and scalable. In addition, we are stepping into 5G Era and the 6G mobile communication is impending, machine/deep learning methods suitable and efficient for the generation and processing of the 5G or 6G big data during construction and maintenances of geoengineering such as tunnels and pipelines are essential, as well as their economic potential is proposed. On the other hand, big data has its own lifecycle, with stages from data acquisition to distribution and archiving etc. Consequently, there are opportunities to leverage machine learning within each stage of the data lifecycle, and they can be summarized into five categories: (1) improving data utilization, (2) automating operations, (3) improving service levels, (4) providing innovative services, and 5) utilizing existing information stewardship skills to curate labeled data. To take advantage of these opportunities, well-curated, large-scale training data sets need to be constructed, as proposed in trend item one.

(7) Based on the principle of "concrete analysis on specific problems", for geo engineering of various complexities, different machine/deep learning and optimization should be adopted. For geo engineering of multivariate big data with noises, algorithms of error correction and considerable robustness is desirable. While for problems of limited features with incomplete inputs, the choice should be quite different. Based on (Phoon 2019), the MUSIC-X characteristics

of geotechnical data (MUSIC-X: multivariate, uncertain and unique, sparse, incomplete, and potentially corrupted, with X denoting the time and/or spatial dimension) decided that the adoption of the methods should not be blind. In other words, optimal results can never be achieved by blindly applying machine-learning models.

(8) Earthquakes, volcanic eruptions, tsunamis or landslides endanger billions of people worldwide. Only if we know more about the underlying physical processes can the effects be minimized. On the one hand, it is needed to develop the next generation of Earth observation systems, and on the other, innovative systems for modelling and reliably interpretation of the big data obtained are essential. From this aspect, the incorporation of new observational quantities and monitoring techniques and the integration of such analyses with ML/DL methods is of course the trend of development. In addition, in the future the developed approaches should sufficiently satisfy the consideration of different time scales ranging from seconds to minutes (alarming such as for landslides, earthquakes), days to weeks (nowcasting such as for volcanic eruptions) to years (forecasting such as for ground subsidence or uplift).

(9) Currently, the assessment or evaluation of geotechnical performances/responses are based on the single regression or classification values, either from numeric numbers or images. More benefits including the improved accuracy or robustness can be obtained via the combined use of numeric values, images or even the audio information. For example, In the process of field geological exploration, geological engineers generally make a preliminary judgment on the lithology classification through the information of rock surface texture, color and hammering audios. Based on expert experience and intelligent mode, a DL and intelligent recognition analysis method for rock lithology classification can be therefore proposed by using coupled rock images and hammering audios.

(10) Another trend in application areas of geoengineering and geosciences is the visual computing, which handles with images and 3D models, i.e. computer graphics, image processing, visualization, computer vision, virtual and augmented reality, video processing, but also includes aspects of pattern recognition, human computer interaction, machine learning and digital libraries. The core challenges are the acquisition, processing, analysis and rendering of visual information (mainly images and video). (Mohammady et al. 2019; Wang et al. 2020) developed CNN EXPLAINER to help users more easily understand the inner workings of CNNs, which can run locally in users' web browsers without the need for installation or specialized hardware, broadening the public's education access to modern deep learning techniques. Recently Stefan Sietzen successfully realized the CNN visualization via Unity 3D and lots of Code (https://vimeo.com/stefsietz).

(11) Although many machine learning methods such as ANN were criticized earlier as a black box engine where the decision-making process is uninterpretable, recent neurosymbolic AI has provided the solution for this obstacle. Moreover, the greater involvement of expert tagging and annotation of the collected data through crowdsourcing, integrative workspaces, and thick data analytics has broadened the availability of good training data sets for solving diverse geoengineering and geoscience problems. Afterall, wide adoption of machine learning in these two areas would be further facilitated only if models become more transparent and interpretable: their parameters and feature rankings should have a minimal physical interpretation, and the model should be reducible to or explainable by a set of rules, descriptors and relations.

(12) Last but not the least, it is essential to regulate the standards for data collection, testing and analysis in establishing the systematic criterion of original data acquisition. In particular, primary data in engineering practice is required to be in conformity with the uniform management standards, which needs to be mandatory by the official management departments and authorities (data privacy). Otherwise, the research progress will be stagnated due to absence of sufficient data samples for machine learning regardless of great efforts in algorithm development. Consequently, engineering application of ML, DL and optimization is not only a scientific issue, but also a management and system topic requiring cooperation of multi parties involved.

References

Mohammady M, Pourghasemi HR, Amiri M (2019) Land subsidence susceptibility assessment using random forest machine learning algorithm. Environ Earth Sci 78. https://doi.org/10.1007/s12665-019-8518-3

Phoon KK (2019) The story of statistics in geotechnical engineering. 14:3–25. https://doi.org/10.1080/17499518.2019.1700423

Wang ZJ, Turko R, Shaikh O et al (2020) CNN Explainer: learning convolutional neural networks with interactive visualization. IEEE Trans vis Comput Graph 27:1396–1406. https://doi.org/10.1109/tvcg.2020.3030418